# Communications
# in Computer and Information Science     **1445**

More information about this series at http://www.springer.com/series/7899

Dmitrii Rodionov · Tatiana Kudryavtseva ·
Angi Skhvediani · Mohammed Ali Berawi (Eds.)

# Innovations in Digital Economy

Second International Scientific Conference, SPBPU IDE 2020
St. Petersburg, Russia, October 22–23, 2020
Revised Selected Papers

Springer

*Editors*
Dmitrii Rodionov ⓘ
Peter the Great St. Petersburg Polytechnic
University
St. Petersburg, Russia

Tatiana Kudryavtseva ⓘ
Peter the Great St. Petersburg Polytechnic
University
St. Petersburg, Russia

Angi Skhvediani ⓘ
Peter the Great St. Petersburg Polytechnic
University
St. Petersburg, Russia

Mohammed Ali Berawi
University of Indonesia
Depok, Indonesia

ISSN 1865-0929           ISSN 1865-0937  (electronic)
Communications in Computer and Information Science
ISBN 978-3-030-84844-6        ISBN 978-3-030-84845-3  (eBook)
https://doi.org/10.1007/978-3-030-84845-3

This Springer imprint is published by the registered company Springer Nature Switzerland AG
The registered company address is: Gewerbestrasse 11, 6330 Cham, Switzerland

# Preface

The International Scientific Conference on Innovations in Digital Economy, SPBPU IDE 2020, was held at the Peter the Great St. Petersburg Polytechnic University, St. Petersburg, Russia, during October 22–23, 2020. This conference was conducted jointly by the Graduate School of Industrial Economics (GSIE) at the Peter the Great St. Petersburg Polytechnic University (SPbPU) and the Center for Sustainable Infrastructure Development (CSID) at the Universitas Indonesia (UI).

The SPBPU IDE 2020 conference brought together experts from academia and industry to uncover challenges and solutions regarding the digital transformation of economic systems. As it took place during the COVID-19 pandemic, our conference was conducted online to minimise the risk of illness spreading among our participants. Nevertheless, we tried to create an atmosphere which facilitated communication and knowledge exchange among our participants on the topic of the digital economy. During the conference, we discussed both purely scientific and applied research results.

The plenary meeting's agenda included presentations by key conference speakers. One of these was Dirk Nitzsche, Senior Lecturer at Cass Business School, who presented a report on the development prospects of the fintech industry, focusing on an overview of the opportunities and threats as well as the effectiveness of the implementation of modern technologies in the financial sector. Tatiana Kudryavtseva, Associate Professor at the Graduate School of Engineering and Economics at the Institute of Industrial Management, Economics and Trade, SPbPU, and Angi Skhvediani, a research assistant at the same institute, presented a case study on the implementation of modern technology to improve the master's program in digital economy and business analytics at the Graduate School of Industrial Economics.

Several special events took place during the conference. The first event was organised by the editorial board of the International Journal of Technology. Mohammed Ali Berawi, Associate Professor and Director of the Center for Sustainable Infrastructure Development (CSID) at the Universitas Indonesia, and Managing Editor Nyoman Suwartha gave a special lecture entitled 'Writing and Publishing Scientific Articles', focusing on the preparation, writing and publication of journal articles. The second event was a round-table discussion entitled 'End-to-end Technologies in Industry' organised by the Industrial Systems of Streaming Data Processing Laboratory of the National Technological Initiative (NTI) SPbPU Center with the participation of its industrial and technological partners. The third event was a round-table discussion held to explore the topic of cross-border cooperation between the regions of St. Petersburg, Leningrad Oblast and south-eastern Finland using digital tools. The Inclusive Cross-border Business Networking of Tomorrow (INCROBB) project was launched by the Lappeenranta University of Technology as a leading partner, along with the Graduate School of Industrial Economics (GSIE) at the Peter the Great St. Petersburg Polytechnic University (SPbPU), the St. Petersburg Chamber of Commerce and Industry, Saimaan-Virta and Etelä-Karjalan Yrittäjät ry. During this event, participants discussed problems with, and ways to facilitate, cross-border cooperation. One of the main suggestions was the development of a website where freelancers and companies from different countries can help each other to establish cross-border cooperation.

There were 117 submissions to the conference in total. All submitted papers passed a four-stage review process. At the first stage, all papers were evaluated and reviewed by the conference or Program Committee co-chairs. At the second stage, papers were evaluated and reviewed by at least two reviewers or a conference Program Committee member. At the third stage, we conducted technical reviews and checked papers for plagiarism, mastery of the English language and overall structure. This resulted in a pool of 63 high-quality papers presenting the best practices and scientific results in the conference topics. Out of these papers, 14 were recommended for publishing in CCIS after additional review and significant extension. Nine papers successfully passed this additional review process and were accepted for inclusion in the CCIS proceedings. These papers were divided into three sections.

For the first section, 'Economic efficiency and social consequences of digital innovations implementation', three papers were selected. The first paper in this section, written by Juaneda-Ayensa, E., Clavel San Emeterio, M., Cirilo-Jordan, S. and González-Menorca, L., presents research dedicated to the unified theory of acceptance and use of social apps (UTAU-SA). The authors discuss the role of technology in the promotion of recycling behaviour. They identify that the key determinants of behavioural intention are, in order of importance, impact awareness, desire for notoriety and facilitating conditions. In addition, they find that the intention to use the application has a direct and positive effect on the intention to recycle. The second paper in this section, written by Mohammed Ali Berawi, Mohamad Khaerun Zuhry Radjilun and Mustika Sari, is dedicated to developing a blockchain-based crowdfunding model for property investment. The authors propose the concept of a crowdfunding platform which utilises blockchain technology. This will help to improve the accountability and transparency of the crowdfunding process. The third paper in this section, written by Kirill Sarachuk, Magdalena Missler-Behr and Adrian Hellebrand, discusses the role of ultra-high-speed broadband internet in firm creation in Germany. The authors find a positive and significant relationship between basic broadband availability and firm creation, while ultra-high-speed connections have a weak and negative influence on new business formation.

For the second section, 'Industrial, service and agricultural digitalisation', three papers were selected. The first paper in this section, written by Nadezhda Kvasha, Ekaterina Malevskaya-Malevich and Svetlana Kornilova, discusses the issues of information modelling technology as the integrating basis of investment process development. The second paper in this section, written by Mariya Volik and Tatyana Kopysheva, reveals the peculiarities of the business-process automation of client interaction in a company based on Bitrix24. The researchers use the example of a software development and implementation company to demonstrate the possible effects of the automation of customer interactions. The third paper in this section, written by Tatina Verevka, Andrei Mirolyubov and Juho Makio, discusses opportunities for and barriers to using big data technologies in the metallurgical industry. The authors discuss possible positive effects of the implementation of machine-learning technologies and predictive analytics tools based on big data platforms.

For the third section, 'End-to-end digital technologies in industry', three papers were selected. The first paper in this section, written by Valeria Rakova, Marina Bolsunovskaya, Arseny Zorin, Vladimir Fedorov and Yuliya Novikova, explores the use of digitalisation to improve population well-being in the arctic area. The authors present a project for the development of the information complex, which can improve monitoring and decision making in the spheres of sanitation and epidemiology for the population of the Russian Arctic. The second paper in this section, written by Nikolay Lomakin, Anastasia Kulachinskaya, Uranchimeg Tudevdagva, Elena Radionova and Natalya Mogharbel, presents an AI-system for predicting Russia's GDP volume based on dynamics of factors in the transport sector. The authors use a perceptron AI system that consists of an input layer and two hidden layers. This system allows them to predict the Russian GDP with an error margin of 0.0027%. The third paper in this section, written by Vasilyanov Georgiy and Vassiliev Alexei, discusses the usage of embedded virtual modelling for the development of a symmetrical walking robotic platform.

We want to thank our keynote speakers, panellists and authors, who contributed to the conference and made it possible by submitting and later revising their work according to the comments provided by the reviewers. We are also grateful to the members of the Program Committee for providing valuable and profound reviews.

We hope that the conference will become an annual event to share recent developments in the sphere of the digital economy.

July 2021

Dmitrii Rodionov
Tatiana Kudryavtseva
Mohammed Ali Berawi
Angi Skhvediani

# Organization

## Conference Chairs

| | |
|---|---|
| Vladimir Schepinin | Peter the Great St. Petersburg Polytechnic University, Russia |
| Hendri Budiono | Universitas Indonesia, Indonesia |
| Dmitrii Rodionov | Peter the Great St. Petersburg Polytechnic University, Russia |
| Mohammed Ali Berawi | Universitas Indonesia, Indonesia |

## Program Committee Chairs

| | |
|---|---|
| Irina Rudskaya | Peter the Great St. Petersburg Polytechnic University, Russia |
| Tatiana Kudryavtseva | Peter the Great St. Petersburg Polytechnic University, Russia |

## Program Committee

| | |
|---|---|
| Alexey Bataev | Peter the Great St. Petersburg Polytechnic University, Russia |
| Anderson Catapan | Universidade Tecnologica Federal do Parana, Brazil |
| Andrea Caputo | University of Lincoln, UK |
| Andrey Zaytsev | Peter the Great St. Petersburg Polytechnic University, Russia |
| Angela Mottaeva | Moscow State University of Civil Engineering, Russia |
| Anis Alamshoyev | Tajik State Finance and Economics University, Tajikistan |
| Anna Tanina | Peter the Great St. Petersburg Polytechnic University, Russia |
| Chen Fangruo | Shanghai Jiao Tong University, China |
| Cristina Sousa | Universidade Portucalense, Portugal |
| Diego Pacheco | Universidade Federal do Rio Grande do Sul, Brazil |
| Antonio Petruzzelli | Polytechnic University of Bari, Italy |
| Ekaterina Abushova | Peter the Great St. Petersburg Polytechnic University, Russia |
| Ekaterina Koroleva | Peter the Great St. Petersburg Polytechnic University, Russia |
| Elena Rytovater | Peter the Great St. Petersburg Polytechnic University, Russia |
| Elena Sharafanova | St. Petersburg State University of Economics, Russia |

## Organizing Committee Chair

Angi Skhvediani                Peter the Great St. Petersburg Polytechnic University, Russia

## Organizing Committee

Anastasia Kulachinskaya        Peter the Great St. Petersburg Polytechnic University, Russia

Darya Kryzhko                  Peter the Great St. Petersburg Polytechnic University, Russia

Natalia Abramchikova           Peter the Great St. Petersburg Polytechnic University, Russia

# Contents

# Economic Efficiency and Social Consequences of Digital Innovations Implementation

# Unified Theory of Acceptance and Use of Social Apps: (UTAU-SA): The Role of Technology in the Promotion of Recycling Behavior

Emma Juaneda-Ayensa[✉] ⓘ, Mónica Clavel San Emeterio ⓘ,
Stephania Cirilo-Jordan, and Leonor González-Menorca ⓘ

Business and Economics Department, University of La Rioja, C/ La Cigüeña 60,
26004 Logroño, La Rioja, Spain
emma.juaneda@unirioja.es

**Abstract.** One of the biggest global problems is the amount of waste discarded and its negative impact. Recycling behavior is a key element in solving this problem, and new information and communication technologies can be a means to promote this desired pindrosocial behavior. The objective of this research is to identify the key factors that condition the acceptance and use of a mobile application to promote recycling behaviors. The present study with a sample of 1.200 Spanish individuals, explain intention to use the mobile application through a new application of the UTAUT model, adding two new constructs, impact awareness and desire for notoriety, and how this intention to use this new application may influence individual's intention to recycle. Results have been analyzed with variance-based structural equation modelling (PLS). The model has a $R^2$ of 0.703 for Intention to use mobile app and 0.378 for recycling intention, and a predictive capacity $Q^2$ of 0.557 and 0.298 respectively. The results indicate that: 1) the key determinants of behavioral intention are, in order of importance, Impact awareness (=0.541), Desire for notoriety (=0.259) and Facilitating conditions (=0.187); 2) there is a direct and positive effect of the intention to use the application on the intention to recycle (=0.615). The results contribute to understand how technology should be designed to promote prosocial behavior.

**Keywords:** Recycling behavior · Unified theory of acceptance and use of technology (UTAUT) · Social apps · Facilitating conditions · Impact awareness · Desire for notoriety

## 1  Introduction

One of the biggest global problems is the amount of waste discarded. This has negative environmental, economic, health and social impacts [1]. The high cost of waste management and decontamination, the wastage of raw materials that could be reused, and the restoration of contaminated areas contribute to the economic problem. Of the waste generated globally only 13.5% is recycled, and only 5.5% is used for compost;

© Springer Nature Switzerland AG 2021
D. Rodionov et al. (Eds.): SPBPU IDE 2020, CCIS 1445, pp. 3–22, 2021.
https://doi.org/10.1007/978-3-030-84845-3_1

thus, more than half of the waste generated globally has no proper management. In low-income countries, only 4% of all waste generated is recycled [2].

The circumstances arising from the 2020 COVID-19 pandemic add one more challenge to the scenario. The enhanced demand for single-use personal protective equipment (PPE) for health sector workers, mandatory mask use, changes in lifestyles and consumption, and the general perception of the hygienic superiority of single-use plastics have caused behavioral changes that have increased the use of these materials [3, 4]. Thus, while plastic pollution was described as a global crisis by the United Nations in 2017 [5], the pandemic has increased the complexities of managing plastic waste, and highlighted the vulnerability of populations worldwide to contamination [4]. In addition, the increase in the use of single-use plastics has made plastic waste management more difficult; this paradigm has caused the disposal of more than 40,000 kg of plastic into the natural world [6].

Governments, NGOs and companies are trying to improve recycling management systems by creating new techniques, modifying legislation and adopting new concepts, such as the circular economy, all with the aim of improving waste management. In this joint effort to improve waste management systems measures are being undertaken to improve their technical efficiency and, from the human viewpoint, to promote more sustainable behaviors and environmental commitment. The effectiveness of waste collection services depends to a large extent on consumers' attitudes and intentions [7–9]. The identification of the factors that determine recycling behaviors is a growing research area [10]. The literature features many studies on recycling and its predictors, but there has been no comprehensive examination of the combination of factors that determine pro-environmental behaviors [11] and, as yet, the literature does not adequately explain consumer intention to participate in recycling [12].

The rapid development of information and communication technologies (ICTs) is creating opportunities for innovation in waste recycling systems [13], and has prompted a considerable increase in research into ICTs and sustainability [14–16]. Information and communication technologies offer solutions that can solve, at least partially, waste-related problems by using IoT concepts to design efficient waste-collection solutions [17], and by using digital information systems to disseminate waste classification and recycling information on social networks to promote pro-environmental behaviors [18]. However, although internet-based systems can be effective in maintaining the interest of consumers who already participate in waste sorting and recycling programs, simply providing information on these programs may not be enough to trigger waste recycling behaviors in those who are as yet unaware of the problem [19].

Thus, in this work we focus on how emergent technologies might promote recycling behavior by providing information that encourages the citizenry to participate in better waste management. Specifically, this work focuses on mobile devices, one of the recent technological developments that has had the greatest societal impact. The objective of the study is to identify the factors that influence the adoption of smartphone-based recycling applications to help mobile application designers develop appropriate strategic actions.

To this end, we present the UTAU-SA (unified theory of acceptance and use of social apps) model, a new model based on the unified theory of acceptance and use of new technologies, the UTAUT [20]; two new constructs are added, impact awareness and desire for notoriety. The results of this study contribute to future research that can help developers of smartphone-based recycling applications understand the key determinants of the adoption of recycling behaviors.

The next section of this article reviews the technology adoption literature and presents the corresponding hypotheses. The third part presents the research methodology, the data analysis and the results. Finally, the discussion, the main theoretical contributions towards the design of new mobile device-based social applications, and future research lines, are presented.

## 2 Literature Review

Mobile devices, in recent years, have become essential lifestyle tools [21]. During the last decade, internet-accessible smartphones have become very widely used [22–26], and are forecast to reach 5.1 billion users in 2020 [27]. This trend has made it possible to develop mobile applications that promote waste classification and recycling [19]. However, as Hoehle et al. [26] argued, many mobile applications have not been as successful as expected and application designers need to understand that different people have different needs.

The rapid development of information and communication technologies (ICT) has created opportunities for innovation in public service provision, and in concrete, in waste recycling systems [13]. In the mid-1990s, public administrations began to have high hopes for the development potential of electronic channels, encouraging their use for both information and service delivery [27]. The expansion and widespread use of ICTs, together with the fact that habit is one of key factor to accept a new technology [29] the reality reflects that citizens do not always prefer the use of new technologies to interact with public administration for the provision of public services [30]. This may be because the designers of new applications and the managers of public services fail to understand the interests of citizens in adopting a new technological approach to interacting with the administration and promoting benefits for the actors involved.

Smartphone-based recycling applications are relatively new and their adoption is in an incipient stage. Mobile phone technology adoption research has mainly applied the UTAUT model to fields such as application-based learning [31], shopping [32], payment and banking [33, 34], (mobile) wallet [35] and fitness [36]; and, while there has been a considerable increase in research into ICTs and sustainability [14–16, 37], there is still a need to identify the capacity of these new technologies to promote socially responsible behaviors and commitment to sustainable development.

Our research framework is based on the unified theory of acceptance and use of technology (UTAUT) [20]. The UTAUT model identifies the key elements in the individual´s acceptance and use of a new information technology, assessing the likelihood of success for new technology introductions and helps to understand the drivers of acceptance in order to proactively design interventions. The UTAUT model consider four construct as direct determinants of user acceptance and use behavior: performance expectancy, effort expectancy, social influence, and facilitating conditions. The first version of this model was defined to be applied in a working context. Almost a decade later, the authors defined a new version of the model (UTAUT2) [29] to broaden the scope of application to use of technology in a consumer context, adding three elements to the model:

hedonic motivation, price value, and habit. We selected the UTAUT model because it provides an explanation for the acceptance and use of ICT in the context of a desirable civic behavior, here, recycling. Due to its greater predictive capacity the UTAUT is considered to be more integrative than previous technology adoption theories [38]. Other factors were taken into account in this choice. The UTAUT: 1) can be applied to different technologies and contexts [29]; 2) was created in relation to the use of mobile phones; 3) can be applied with added variables and integrated with other models [39]. Furthermore, Venkatesh et al. [40]. recently recommended introducing new endogenous and exogenous constructs to the UTAUT/UTAUT2 models to increase their explanatory power [36].

## 2.1 Model and Hypothesis Development

Since the first studies into technology adoption it has been assumed that users follow rational decision-making processes based on the cost-benefit binomial. Therefore, based on the theory of reasoned action, the technology acceptance model (TAM) [41] argued that the two primary characteristics necessary to accept an information system are perceived ease of use and perceived utility [42]. The TAM model has attracted various criticisms, which prompted the development of other models, such as the UTAUT [20]. As previously noted, the UTAUT model proposed key factors for the acceptance of new technologies in the work context. Subsequently, the authors expanded the UTAUT to the context of consumer products, which gave rise to the UTAUT2 [29]. Both models have been extensively employed. These substantial replications, applications, and extensions/integrations of UTAUT/UTAUT2 have been valuable in expanding understanding of technology adoption and extending the theoretical boundaries of the theory. However, while our literature review revealed previous works in the context of new applications, we found few that applied the model in the context of public service. Osborne [29] points to four characteristics that differentiate public services from private services. First, they are services that are not intended to expand market share or to build user loyalty; rather, loyalty would be a signal of system failure. of public services. Second, the reality of unwilling or coerced customers is a marked element of public services (e.g. in the prison service or child protection services). The third characteristic would be the existence of multiple stakeholders with different, and in some cases conflicting, objectives, requiring a higher level of negotiation in the definition of the service to promote the satisfaction of all parties involved. And finally, *public service users also inhabit the dual role of being both the users of public services and citizens who may have a broader, societal interest, in the outcomes of public services.* Waste management, and recycling, are a service in that the objective is not to promote a greater demand for the service, but rather that the ultimate goal would be that there would be no need to provide the service because no waste is generated. The interests of different parties such as administrations, producers and consumers also converge directly. Finally, it is a service whose result falls on the environment and on humanity in general, since waste generation has become a global problem. Waste management, as a public service system [43], is strongly dependent on the co-production tasks implemented by citizens.

Thus, there is still a need for a systematic examination and theorizing of the salient factors that might apply to a consumer technology use context [29]; thus, to open this new context we identified the key factors of the UTAUT model and added 2 new factors based on the use of social apps.

In a contribution, therefore, to the UTAUT literature, we added two new exogenous constructs to the original theoretical model developed by Venkatesh et al. [20], that is, impact awareness and desire for notoriety.

**Performance Expectancy.** As defined in the UTAUT model, Performance expectancy is the degree to which an individual believes that using the system will help him or her to attain gains in job performance [20]. This variable reflects individuals' perceptions of a system's operational performance, such as effectiveness and convenience. The five constructs from the different models that pertain to performance expectancy are perceived usefulness (TAM/TAM2 and C-TAM-TPB), extrinsic motivation (MM), job-fit (MPCU), relative advantage (IDT), and outcome expectations (SCT) [20]. This construct is one of the most important factors in the fostering of behavioral intentions and behaviors [29]. Gao et al. [7] argued that when online-based recycling services prove themselves useful by providing waste disposal benefits, the intention to use these types of service increases. In the case of the mobile application, the aim would be to make it easier for the individual to recycle by providing useful information on how to recycle and how to do it correctly. This includes information on the types of waste, whether they are recyclable and, if so, in what type of recycle bin they should be deposited. If the apps are perceived as useful in this sense, they will be a key element for the individual to want to use the app.

The present study measures the relationship between performance expectancy for a mobile application and intention to use it, thus:

H1: The performance expectancy for a mobile application has a positive effect on intention to use the application.

**Effort Expectancy.** Effort expectancy has been defined as the user's perception of how easy a system is to use [29]. This variable is considered fundamental in the adoption and use of technology [44]. Gao et al. [7] argued that if users perceive that it is easy to use a particular technology to achieve their goals, and using it does not require much effort, they will have an expectation of high performance. Thus, this variable reflects the user's perception of how difficult it will be to use a recycling-based mobile application: therefore, the following hypothesis is proposed:

H2: The users' effort expectancy of using a mobile application has a positive effect on intention to use the mobile application.

**Facilitating Conditions.** Venkatesh et al. [29] defined facilitating conditions as the user's perception of the resources and support available to help him/her carry out a behavior. These authors also argued that facilitating conditions determine the use of technologies. Some studies have proposed that this variable is very similar to the perceived behavioral control variable of Ajzen's [45] TPB model, given that it shows the effect of users' knowledge, skills and resources [7]. In this sense, if users have the necessary devices, compatible operating systems, knowledge and support services, they would be willing to use the mobile application. Thus, the following hypothesis is proposed:

H3: The facilitating conditions that support a mobile application have a positive effect on intention to use the mobile application.

**Social Influence.** This variable relates to the user's perceptions of the importance that influential groups, for example, his/her family, give to him/her using a particular technology [29]. Social influence, today, affects actual technology use [44]. Isaac et al. [44] also argued that social influence has a positive impact on internet use. In the context of new apps, there is a consensus in the mobile commerce literature that the social features of new mobile technologies (e.g. online ratings, online reviews and online tracking) play an important role [46].

In the context of this research, social influence on the use of recycling-based mobile applications is defined as the importance that the user's reference groups give to whether (s)he should use a recycling app. Thus:

H4: Social influence has a positive effect on intention to use a recycling-based mobile application.

**Impact Awareness.** Many people habitually perform pro-environmental behaviors, that is, behaviors that mitigate the harmful impact of human activities on the environment [47]. Consumers have been shown to have environmental awareness when buying, or using, certain types of products, but this is not always transferred when performing behaviors, which raises the concept of the consumer's perception of effectiveness, a variable that is important for explaining environmentally conscious consumer behavior. Impact awareness relates to the judgements people make about their ability to influence environmental resource issues; it is based on the concept that when consumers can act to reduce pollution, they take account of the environmental impact of their actions [48].

Thus, for this research, the impact awareness variable was created. The variable is intended to measure whether it is important for users when they start using the application: to understand the environmental and social impact of their actions; to understand the impact that their actions have on their own environment; to know if participating in a recycling program will make them feel more environmentally friendly. This variable is based on the term perceived consumer effectiveness (PCE) [48] and we have adapted to our purpose, an individual's intention to use the mobile application if it allows them to monitor their recycling behavior and be aware of the impact he/she has on the environment.

The related hypothesis is as follows:

H5: The impact awareness of recycling has a positive effect on intention to use the recycling-based mobile application.

**Desire for Notoriety.** The pro-environmental behavior that individuals may exhibit is based on the recognition of certain individual values. Convictions predict behavioral intentions, which provoke real behaviors and begin a causal process through which environmental beliefs become evident [47].

Kang et al. [48] introduced and explained the concept of perceived personal relevance, which they defined as the individual's belief that a certain behavior is associated with his or her lifestyle, values and self-image. When individuals have self-images that they are environmentally responsible they tend to show pro-environmental attitudes. Wu et al. [49] argued that consumers share their values through networking on social media. For the purposes of this research we propose the variable desire for notoriety; its purpose is to identify if using the mobile application allows users: to present themselves on social networks as environmentally responsible; to show to others the lifestyle they want to lead; to let others see them as they want to be seen, that is, as environmentally responsible. Thus, the following hypothesis is proposed:

H6: The desire for notoriety has a positive effect on intention to use the recycling-based mobile application.

**The Impact of Technology on Recycling Behavior.** Mobile technologies have become part of people's everyday lives; their reach has now extended to incorporate activities such as recycling [50]. From the business viewpoint, technologies have been used to develop omnichannel strategies and enhance consumer-company interactions and encourage consumer purchasing behaviors [51]. These omnichannel strategies have not been used to develop pro-environmental behaviors such as recycling, and there is limited literature on the effects of new technologies, for example apps, on recycling behaviors; however some studies have examined how technology has promoted changes in pro-environmental behaviors. For example, many empirical works have substantiated the effects of television (old media) in predicting attitudes toward environmental issues [52], the effects of the news media on pro-environmental behavior [53], and how e-commerce has been used to promote electronic (e-waste) recovery [50]. Thus, we propose the following hypothesis:

H7: Intention to use a recycling-based mobile application has a positive effect on recycling intention.

## 3 Material and Methods

### 3.1 Measurement

The measurement scales were adapted from the previous literature, as we show at Table 1. The UTAUT items performance expectancy, effort expectancy, facilitating conditions and social influence were adapted from Venkatesh et al. [20, 29], and Chen [54], to the particular context of the acceptance and use of recycling-based mobile applications; the measurements of impact awareness and desire for notoriety were adapted from Kang et al. [48], and of intention to recycle from Kumar [12]. The total number of items measured, using 11-point Likert-type scales, was 29 (0 totally disagree, and 10 totally agree).

While the measurement scales have been used in various fields, they have not been applied to recycling: thus, they were first translated from English into Spanish using a back-translation method, whereby one person translated the items into Spanish; they were then adapted to the recycling field. Thereafter, an official, native-English speaking translator translated them back into English; this made it possible to check for any misunderstandings or misspellings resulting from the translation [55]. In addition, we conducted a pretest with 50 participants to ensure the comprehensibility of the questions.

**Table 1.** Measurement scales

| Var. | Adaptation description | [20] | [29] | [54] | [48] | [12] |
|---|---|---|---|---|---|---|
| Performance expectancy | The mobile application must be useful for recycling. | √ | √ | | | |
| | The mobile application should save me time in looking for information about the recycling of certain containers. | √ | √ | | | |
| | The mobile application will help me to recycle waste properly. | √ | √ | | | |
| | The mobile application will allow me to understand the impact of my recycling on the environment. | √ | √ | | | |
| Effort expectancy | The mobile application must be easy to learn to use. | √ | √ | | | |
| | The mobile application should allow intuitive interaction. | √ | √ | | | |
| | The mobile application should be easy for me to use. | √ | √ | | | |
| Facilitating conditions | I have the necessary devices to use the app. | √ | √ | | | |
| | I have the necessary knowledge to use the app | √ | √ | | | |
| | I have a compatible operating system and available capacity. | √ | √ | | | |
| | I find it easy to use the new app. | √ | √ | | | |
| | The mobile application should have customer service available in case a problem arises. | √ | √ | | | |
| Social influence | People who influence my behavior would recommend I use the app. | √ | √ | | | √ |
| | People whose opinion I value would recommend the app to me. | | √ | | | √ |
| | If competent authorities recommend its use. | √ | | | | |
| | People Important to me have downloaded and are using the app. | √ | √ | | | √ |
| Impact awareness | I would use the application because I consider it important to know the environmental and social impact of my actions. | | | | √ | |
| | I would use the application because I would like to understand my impact on the environment. | | | | √ | |
| | Using the app and participating in a recycling program would make me evaluate my impact on, and commitment to, the environment. | | | √ | √ | |
| Desire for notoriety | Using the mobile app would demonstrate on social networks that I am an environmentally responsible person. | | | | √ | |
| | Participating in recycling programs helps me demonstrate the kind of life I want to lead. | | | | √ | |
| | Participating in an app-based recycling program allows others to see me as I would like them to see me. | | | | √ | |
| Intention of using the mobile | I intend to download the mobile app. | √ | √ | √ | | |
| | I intend to use the mobile app to participate in a recycling program. | √ | √ | √ | | |
| | I anticipate that I will continue with the recycling program. | √ | √ | √ | | |
| Intention to Recycle | I intend to recycle containers and cans. | √ | √ | | | |
| | I believe that in the future I will recycle containers and cans. | √ | √ | | | √ |
| | I believe that recycling containers and cans will become normalized in my daily life. | √ | √ | | | √ |
| | I intend to make a habit of recycling containers and cans. | √ | √ | | | √ |

## 3.2 Data Collection and Sample

To collect the data we designed an online survey; this was administered to a Spanish internet panel by the CINT company. The final sample consisted of 1200 Spanish individuals. As to the sample characteristics, the participants were equally distributed in terms of age and gender; the field work was carried out in September 2020.

## 3.3 Data Analysis

To analyze the data set structural equation modeling (SEM), considered a very robust and powerful statistical tool in various disciplines [56], was used. In undertaking SEM, it is advised to use the PLS-SEM approach when: (1) the research is exploratory; (2) the focus is on predicting phenomena, rather than understanding the relationships between the phenomena; and (3) the model is particularly complex [56], as in our case. Furthermore, PLS-SEM may have higher statistical predictive power than CB-SEM [56], and most previous research has used PLS-SEM in studies into recycling and the TPB [57]. For these reasons we chose PLS-SEM (with SmartPLS 3.0).

# 4 Results

## 4.1 Measurement Model

The analysis of the measurement scale did not return favorable discriminant validity results for the variables effort expectancy and performance expectancy, due to their high degree of correlation. Using IBM SPSS 24.0 an exploratory factor analysis was performed for the set of items of both constructs, which resulted in a single factor that explained 74.45 of variance; thus, we created a "cognitive evaluation" construct to group the items of expectancy and performance expectancy, and reformulated hypotheses H1 and H2 into a single hypothesis:

H1: The user's cognitive evaluation of a recycling-based mobile application has a positive effect on his/her intention to use the application.

This correlation may be due to the fact that both variables are based on behavioral psychology models [51], among them Ajzen's TPB [45]. For their part, these factors are based on utilitarian motivational factors [41], which justifies the decision to merge the two constructs to create a new construct. Our review of the literature suggested a further explanation for the high correlation; it may be because users have, today, a high degree of knowledge of mobile applications, and easy access to mobile devices, as we note in the review. Following the modification of the analytical model, the reliability and convergent validity of the measurement scales were assessed through their Cronbach's alphas, the composite reliability index and the average variance extracted (AVE). The constructs' internal consistencies were high, as can be seen at Table 2. The Cronbach's alphas were greater than the recommended value of 0.7 [58]. The composite reliability results were all above 0.6 [59], and the AVEs were higher than 0.5 [60, 61]. The significance of the loadings was determined through the bootstrap resampling procedure (5000 subsamples of the original sample size). All the items were significantly related to each of their factors ($p < 0.01$), and all loadings were above 0.7 [61].

**Table 2.** Reliability and convergent validity

|  | Cronbach's alpha | CR | AVE |
|---|---|---|---|
| Cognitive evaluation | 0.943 | 0.943 | 0.702 |
| Desire for notoriety | 0.892 | 0.891 | 0.733 |
| Facilitating conditions | 0.879 | 0.874 | 0.581 |
| IU Social mobile app | 0.936 | 0.936 | 0.829 |
| Impact awareness | 0.939 | 0.939 | 0.838 |
| Recycling intention | 0.956 | 0.956 | 0.844 |
| Social influence | 0.876 | 0.874 | 0.637 |

Discriminant validity was assessed using the Fornell-Larcker criterion and the heterotrait-monotrait ratio (HTMT) of correlations. As can be seen at Table 3, all values of the square roots of the AVEs were greater than the values of the inter-construct correlations, and the heterotrait-monotrait ratio had values <0.9. Based on these outcomes the discriminant validity of the measurement model is confirmed [62].

**Table 3.** Discriminant validity

|  | Cognitive evaluation | Desire for notoriety | Facilitating conditions | IU Social mobile app | Impact awareness | Recycling intention | Social influence |
|---|---|---|---|---|---|---|---|
| Cognitive evaluation | 0.838 | 0.344 | 0.691 | 0.624 | 0.768 | 0.709 | 0.414 |
| Desire for notoriety | 0.354 | 0.856 | 0.317 | 0.603 | 0.529 | 0.272 | 0.679 |
| Facilitating conditions | 0.697 | 0.331 | 0.762 | 0.638 | 0.694 | 0.631 | 0.405 |
| IU Social mobile app | 0.626 | 0.608 | 0.644 | 0.911 | 0.802 | 0.614 | 0.521 |
| Impact awareness | 0.770 | 0.537 | 0.702 | 0.802 | 0.915 | 0.698 | 0.505 |
| Recycling intention | 0.708 | 0.281 | 0.635 | 0.615 | 0.698 | 0.918 | 0.291 |
| Social influence | 0.422 | 0.669 | 0.421 | 0.524 | 0.510 | 0.297 | 0.798 |

Note: The values on the main diagonal are the square roots of the AVEs; the values below the diagonal are the correlations between factors; the values above the diagonal are the HTMT ratios.

## 4.2 Structural Model

The $R^2$ has a value of 0.703 for IU social mobile apps and 0.378 for recycling intention, which indicates that the model explains 70.3% of the variation in intention to use the recycling mobile application and 37.80% of the variance of recycling intention. The $Q^2$ of IU social mobile apps was 0.557, and 0.298 for recycling intention. These results showed that the model has high predictive capacity [63, 64].

As to the results obtained from the structural model to analyze the proposed research model, it was shown that all the ß were significant at $p < 0.01$, except those of the cognitive evaluation and social influence constructs, as shown at Table 4. Thus, four of the six research hypotheses are supported. Hypotheses H1 (ß = −0.013, $p > 0.01$) and H4 (ß = 0.001, $p > 0.01$) were rejected; therefore, we did not find empirical evidence that allows us to affirm that social influence and cognitive evaluation significantly influence intention to use the mobile application. As can be seen, the variable that most directly influences intention to use the application is impact awareness (ß = 0.541, $p < 0.01$). Intention to recycle is strongly influenced by intention to use the mobile application (ß = 0.615, $p < 0.01$) (Fig. 1).

**Table 4.** Structural model results

|  | Path coefficient | STDEV | T | P | 2.5% | 97.5% |
|---|---|---|---|---|---|---|
| Cognitive evaluation -> IU Social mobile app | −0.013 | 0.041 | 0.328 | 0.743 | −0.095 | 0.067 |
| Desire for notoriety -> IU Social mobile app | 0.259 | 0.040 | 6.517 | 0.000 | 0.183 | 0.338 |
| Facilitating conditions -> IU Social mobile app | 0.187 | 0.046 | 4.083 | 0.000 | 0.101 | 0.277 |
| IU Social mobile app -> Recycling intention | 0.615 | 0.030 | 20.842 | 0.000 | 0.556 | 0.672 |
| Impact awareness -> IU Social mobile app | 0.541 | 0.051 | 10.687 | 0.000 | 0.436 | 0.636 |

Note: Bootstrapping 95% bias corrected confidence intervals (based on n = 5000 subsamples).

## 5 Discussion

This research has important implications for sustainable behavior of citizen and their participation in the waste management systems as it sheds new light on a scarcely-studied issue and represents a starting point for further research. The present study is the first to analyze the factors that trigger the use of social mobile application in citizen recycling behavior. Furthermore, the study represents a new perspective in the involvement of individuals in the development of public services. It highlights also the importance of information technologies in changing public services and make them more efficient.

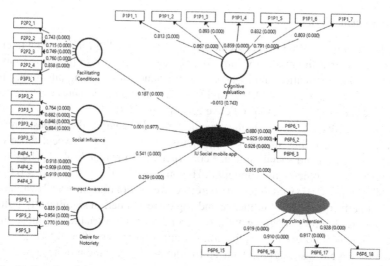

**Fig. 1.** Results of the measurement and structural models: path coefficients (p-values)

In concrete, our article contributes to the research on the conditioning factors of the acceptance and use of new technologies in a different context such as waste management, public services, as a complement to the previous models UTAUT (work context) and UTAUT 2 (consumer context). Few studies have analyzed the antecedents of the acceptance and use of mobile apps that promote recycling behavior from the user perspective [42].

This study presents the UTAU-SA research model, developed from the UTAUT [20], which explains intention to accept a new technology based on the variables effort expectancy, performance expectancy, facilitating conditions and social influence. The original model has been expanded to include two new variables, impact awareness and desire for notoriety [48]. This adaptation is justified by the change in the analysis context, that is, the use of technological applications in the services field, of a public or quasi-public nature, to promote socially responsible behaviors that provide social benefits, in this case, sustainable development.

The evaluation of the model showed that the explanatory power of the variables for intention to use the mobile application is 70% ($R^2 = 0.703$), and this, in turn, has a significant, positive influence, explaining 37% ($R^2 = 0.378$) of recycling intention.

## 5.1 Theoretical Contributions

This study contributes to the academic literature on UTAUT by identifying and re-evaluating the influence of the key aspects that a public goods/services-based mobile application should include to promote environmentally responsible behaviors, specifically recycling behavior.

In the initial evaluation of the measurement model the results for effort expectancy and performance expectancy showed a lack of discriminant validity and, therefore, after the exploratory factor analysis, we redesigned the model and added a new variable, that we called cognitive evaluation, that reflect the user's assessments of expected effort and expected rewards. The results of the EFA were consistent with the cost-benefit paradigm of behavioral decision theory [65]. "People tend to use or not use an application to the extent they believe it will help them perform better. Second, even if potential users believe that a given application is useful, they may, at the same time believe that the system is too hard to use and that the performance benefits of usage are outweighed by the effort of using the application" [41].

The results obtained in the evaluation of the structural model validate the extension of the UTAUT model to predict intention to use a mobile app that promotes recycling behavior. In addition, the visibility of the pro-environmental behavior (i.e. the desire for notoriety) and the awareness of the impact the recycling behavior would have on the environment (impact awareness) are two of the most influential variables in the acceptance and use of the mobile application; facilitating conditions were also found to be influential. The results showed that neither the influence of reference persons (social influence), nor the assessment of perceived effort-performance (cognitive evaluation), had a significant effect on intention to use the mobile application.

Impact awareness was the model construct that most explained intention to use the recycling-based mobile app. In this adaptation of the UTAUT model the main difference between the application of the model in work and consumer contexts is that the beneficiary of the behavior is not the individual, but society in general. From this viewpoint, recycling aims to preserve the environment and reduce the negative impact that environmental degradation has on society. On the other hand, individuals are also influenced by the possibility of being made aware of the environmental and social impacts of participating in recycling programs, and by the feeling of pride generated by this explicit manifestation, which can be compared to the behaviors of volunteer workers engaging in activities that promote the common interest [66]. This result is in line with Kang et al. [48], who argued that the awareness that individuals have of their ability to influence environmental problems is one of the conditioning factors of their behaviors.

Desire for notoriety was the second most important construct in the explanation of intention to use the app. The results indicated that individuals feel the need to make their behavior visible and promote their reputations as people committed to the environment. The theory proposes that individuals believe that certain behaviors are associated with desired lifestyles, that performing these behaviors reinforces their self-image [48], and the public manifestation of these behaviors is related to the management of their reputations on social networks [67]. From the viewpoint of pro-environmental behaviors, individuals act differently in the public and in the private spheres [68]. In this sense, recycling is considered as a manifestation of private behaviors but, through the application, it becomes a behavior in the public sphere; this creates consistency between the individual's beliefs, his/her behaviors and his/her public image. It should be noted that users want to project themselves on social networks as environmentally responsible and as following an admirable lifestyle; this result is consistent with El Marrakchi [67], who highlighted the importance for individuals of improving their public images, especially on social networks.

The third significant variable is the importance of the availability of the means to accept the technology (Facilitating conditions). Information technology and digital marketing researchers have often found that having the means to adopt new technologies has a crucial impact on customers' intentions, and actual use behaviors [21, 46, 69–72].

Although some of the literature has recognized the impact of social influence on people´s attitudes, intentions and behaviors [20, 73], our results showed that this factor does not influence intention to use the app, which is consistent with some other studies [21, 51, 74]. The prior literature posits that collectivist cultures are socially oriented and value group opinions more than individual opinions [75]. However, pro-environmental activities are perceived as a highly personal, and while people around the individual might influence his/her recycling intentions, they will not influence his/her intention to use a recycling-based app. Another reason could be that their use is not obligatory [74]. From a pragmatic viewpoint, it is possible that recycling apps are not, as yet, sufficiently established in Spain to attract strong social influence and, thus, individuals do not perceive social pressure to use recycling-based apps. This leads us to conclude that accepting the application is unrelated to the pressure one feels in one's social environment, and that the wish to influence others is more important, as shown by the results of the variable desire for notoriety. In these circumstances, subjective norms and sense of image (social influence) seem to relate to the individual's perceptions, rather than his/her wish blindly to follow fashion (i.e. it's fashionable to be environmentally responsible), or his/her wish simply to show off in public, or to his/her friends. These results are similar to those of UTAUT model studies in which social factors were not found to influence technology adoption [69, 76–78], and of studies specifically examining mobile application adoption [21, 27, 79–81].

Finally, contrary to previous findings [21, 29, 82], it was shown that the individual's cost-benefit assessment of performing recycling behavior and the effort needed to operate the app (cognitive evaluation, the new variable) did not influence intention to use the app. Our interpretation of these results is that cognitive evaluation does not influence users' intentions to use the application because they do not perceive that it will provide any real, direct benefit in terms of enhancing their recycling performance. In this sense, the app provides an information system that can help its users resolve their doubts, but it is probable they do not perceive this as key because they don´t believe this to be necessary, or are accustomed to consult other sources to achieve clarity. Our results are consistent with previous studies [36, 83, 84]; these authors explained that performance expectancy is not significant with emergent technologies and, thus, potential users are not conscious of their utility. Other works have shown that effort expectancy is not significant [85]. This seems to suggest that making apps easy to operate does not influence users' intention to use them. Thus, the ease of use of an app is not important for consumers when they assess its usefulness.

The present study has some limitations. Our data relate to an extended UTAUT model that analyzes the promotion of a sustainable behavior, in this case recycling, in a voluntary context. Because we adapted the theory to a new context we did not include in our model some variables included in UTAUT2, such as habit and hedonic motivations. In contrast to contexts examined in previous studies, in the UTAU-SA model the proposed context is the use of technology to promote recycling behavior, which provides benefits for the wider society. In this sense, this new context addresses the relationship between intention to use the app (or the technology) and intrinsic motivation-related aspects, but it would also be interesting to incorporate extrinsic motivations into the model to evaluate the importance of rewards in the adoption a new technology developed in a voluntary context for societal purposes. On the other hand (but linked to the previous concept), the incorporation of gamification and hedonic motivational aspects into the application might enhance the explanatory power of the model for audiences reluctant to use the application to improve the environment, but who may use it for entertainment, or to win prizes. This opens a new development line for these technologies that might include aspects such as gamification, evaluations of different types of rewards, and the possibility of understanding more about environmental issues, and of influencing the environment, such as the evaluation of behavioral change over time.

**Acknowledgements.** This research has been made possible thanks to the support of ECOEMBES S.A and funded by European Fund for Regional Development and Economic Development Agency of La Rioja-ADER (IRIS 4.0 project, Grant No. 2017-1-IDD-00057).

# References

1. Smith, N.: Waste Generation and Recycling Indices 2019: Overview and Findings. Verisk Maplecroft, Bath, UK (2019)
2. World Bank, What A Waste Global Database (2018). https://datacatalog.worldbank.org/dataset/what-waste-global-database. Accessed 10 Dec 2020
3. Klemeš, J.J., Van Fan, Y., Tan, R.R., Jiang, P.: Minimising the present and future plastic waste, energy and environmental footprints related to COVID-19. Renew. Sustain. Energy Rev. **127**, 109883 (2020). https://doi.org/10.1016/j.rser.2020.109883
4. Vanapalli, K.R., et al.: Challenges and strategies for effective plastic waste management during and post COVID-19 pandemic. Sci. Total Environ. **750**, 141514 (2020). https://doi.org/10.1016/j.scitotenv.2020.141514
5. UNEP, Actions to address plastic waste: Basel convention (2019). http://www.basel.int/Implementation/Plasticwaste/Overview/tabid/6068/Default.aspx. Accessed 09 Dec 2020
6. WWF, Fondo Mundial para la Naturaleza. https://www.wwf.es/informate/actualidad/?54741/Hay-que-tomar-medidas-decididas-para-frenarlacontaminacion-de-plasticos-de-un-solo-uso. Accessed 01 Sept 2020
7. Gao, S., Shi, J., Guo, H., Kuang, J., Xu, Y.: An empirical study on the adoption of online household e-waste collection services in China. In: Janssen, M., et al. (eds.) I3E 2015. LNCS, vol. 9373, pp. 36–47. Springer, Cham (2015). https://doi.org/10.1007/978-3-319-25013-7_4
8. Król, A., Nowakowski, P., Mrówczyńska, B.: How to improve WEEE management? Novel approach in mobile collection with application of artificial intelligence. Waste Manage. **50**, 222–233 (2016). https://doi.org/10.1016/j.wasman.2016.02.033
9. Wang, W., Qu, Y., Liu, Y., Zhang, Y.: Understanding the barriers for Internet-based e-waste collection system in China. J. Environ. Planning Manage. **63**(4), 629–650 (2020). https://doi.org/10.1080/09640568.2019.1601618
10. Oztekin, C., Teksöz, G., Pamuk, S., Sahin, E., Kilic, D.S.: Gender perspective on the factors predicting recycling behavior: implications from the theory of planned behavior. Waste Manage. **62**, 290–302 (2017). https://doi.org/10.1016/j.wasman.2016.12.036
11. Onel, N., Mukherjee, A.: Why do consumers recycle? A holistic perspective encompassing moral considerations, affective responses, and self-interest motives. Psychol. Market. **34**(10), 956–971 (2017). https://doi.org/10.1002/mar.21035

12. Kumar, A.: Exploring young adults' e-waste recycling behaviour using an extended theory of planned behaviour model: a cross-cultural study. Resour. Conserv. Recycl. **141**, 378–389 (2019). https://doi.org/10.1016/j.resconrec.2018.10.013

13. Zhang, Y., Wu, S., Rasheed, M.I.: Conscientiousness and smartphone recycling intention: the moderating effect of risk perception. Waste Manage. **101**, 116–125 (2020). https://doi.org/10.1016/j.wasman.2019.09.040

14. Artal-Tur, A., Villena-Navarro, M., Alamá-Sabater, L.: The relationship between cultural tourist behaviour and destination sustainability. Anatolia **29**(2), 237–251 (2018). https://doi.org/10.1080/13032917.2017.1414444

15. Bonadonna, A., Giachino, C., Truant, E.: Sustainability and mountain tourism: the millennial's perspective. Sustainability **9**(7), 1219 (2017). https://doi.org/10.3390/su9071219

16. Souza, V.S., de Vasconcelos Marques, S.R.B., Veríssimo, M.: How can gamification contribute to achieve SDGs? J. Hospitality Tourism Technol. **11**(2), 255–276 (2020). https://doi.org/10.1108/JHTT-05-2019-0081

17. Marques, P., et al.: An IoT-based smart cities infrastructure architecture applied to a waste management scenario. Ad Hoc Network. **87**, 200–208 (2019). https://doi.org/10.1016/j.adhoc.2018.12.009

18. Aguiar Castillo, L., Rufo Torres, J., De Saa Pérez, P., Pérez Jiménez, R.: How to encourage recycling behaviour? The case of WasteApp: a gamified mobile application. Sustainability **10**(5), 1544 (2018). https://doi.org/10.3390/su10051544

19. Rasmussen, M.B., Pagels, K.Ø., Ramanujan, D.: Supporting household waste sorting practices by addressing information gaps. J. Comput. Inf. Sci. Eng. **20**(4) (2020). https://doi.org/10.1115/1.4046734

20. Venkatesh, V., Morris, M.G., Davis, G.B., Davis, F.D.: User acceptance of information technology: toward a unified view. MIS Q. **27**(3), 425–478 (2003). https://doi.org/10.2307/30036540

21. Hew, J.J., Lee, V.H., Ooi, K.B., Wei, J.: What catalyses mobile apps usage intention: an empirical analysis. Ind. Manage. Data Syst. **115**(7), 1269–1291 (2015). https://doi.org/10.1108/IMDS-01-2015-0028

22. Wang, S., Barnes, S.J.: An analysis of the potential for mobile auctions in China. Int. J. Mobile Commun. **7**(1), 36–49 (2009). https://doi.org/10.1504/IJMC.2009.021671

23. Brown, S.A., Dennis, A.R., Venkatesh, V.: Predicting collaboration technology use: integrating technology adoption and collaboration research. J. Manage. Inf. Syst. **27**(2), 9–54 (2010). https://doi.org/10.2753/MIS0742-1222270201

24. Hu, P.J.H., Chen, H., Hu, H.F., Larson, C., Butierez, C.: Law enforcement officers' acceptance of advanced e-government technology: a survey study of COPLINK Mobile. Electron. Commer. Res. Appl. **10**(1), 6–16 (2011). https://doi.org/10.1016/j.elerap.2010.06.002

25. Ou, C.X., Davison, R.M.: Interactive or interruptive? Instant messaging at work. Decis. Support Syst. **52**(1), 61–72 (2011). https://doi.org/10.1016/j.dss.2011.05.004

26. Hoehle, H., Zhang, X., Venkatesh, V.: An espoused cultural perspective to understand continued intention to use mobile applications: a four-country study of mobile social media application usability. Eur. J. Inf. Syst. **24**(3), 337–359 (2015). https://doi.org/10.1057/ejis.2014.43

27. Tam, C., Santos, D., Oliveira, T.: Exploring the influential factors of continuance intention to use mobile apps: extending the expectation confirmation model. Inf. Syst. Front. **22**(1), 243–257 (2018). https://doi.org/10.1007/s10796-018-9864-5

28. Ebbers, W.E., Pieterson, W.J., Noordman, H.N.: Electronic government: rethinking channel management strategies. Gov. Inf. Q. **25**(2), 181–201 (2008). https://doi.org/10.1016/j.giq.2006.11.003

29. Venkatesh, V., Thong, J.Y., Xu, X.: Consumer acceptance and use of information technology: extending the unified theory of acceptance and use of technology. MIS Q. **36**(1), 157–178 (2012). https://doi.org/10.2307/41410412

30. Rey-Moreno, M., Medina-Molina, C.: Omnichannel strategy and the distribution of public services in Spain. J. Innov. Knowl. **1**(1), 36–43 (2016). https://doi.org/10.1016/j.jik.2016.01.009

31. Pindeh, N., Suki, N.M., Suki, N.M.: User acceptance on mobile apps as an effective medium to learn Kadazandusun language. Procedia Econ. Finance **37**, 372–378 (2016). https://doi.org/10.1016/S2212-5671(16)30139-3

32. Miladinovic, J., Hong, X.: A study on factors affecting the behavioral intention to use mobile shopping fashion apps in Sweden (2016). https://www.diva-portal.org/smash/get/diva2:933382/FULLTEXT01.pdf. Accessed 11 Dec 2020

33. Kim, C., Mirusmonov, M., Lee, I.: An empirical examination of factors influencing the intention to use mobile payment. Comput. Hum. Behav. **26**(3), 310–322 (2010). https://doi.org/10.1016/j.chb.2009.10.013

34. Baptista, G., Oliveira, T.: Understanding mobile banking: the unified theory of acceptance and use of technology combined with cultural moderators. Comput. Hum. Behav. **50**, 418–430 (2015). https://doi.org/10.1016/j.chb.2015.04.024

35. Madan, K., Yadav, R.: Behavioural intention to adopt mobile wallet: a developing country perspective. J. Indian Bus. Res, **8**(3), 227–244 (2016). https://doi.org/10.1108/JIBR-10-2015-0112

36. Dhiman, N., Arora, N., Dogra, N., Gupta, A.: Consumer adoption of smartphone fitness apps: an extended UTAUT2 perspective. J. Indian Bus. Res. **12**(3), 363–388 (2019). https://doi.org/10.1108/JIBR-05-2018-0158

37. Lerario, A., Di Turi, S.: Sustainable urban tourism: reflections on the need for building-related indicators. Sustainability **10**(6), 1981 (2018). https://doi.org/10.3390/su10061981

38. Mosquera, A., Olarte-Pascual, C., Ayensa, E.J., Murillo, Y.S.: The role of technology in an omnichannel physical store. Span. J. Market. ESIC (2018). https://doi.org/10.1108/SJME-03-2018-008

39. Zhao, Y., Bacao, F.: What factors determining customer continuingly using food delivery apps during 2019 novel coronavirus pandemic period? Int. J. Hospitality Manage. **91**, 102683 (2020). https://doi.org/10.1016/j.ijhm.2020.102683

40. Venkatesh, V., Thong, J.Y., Xu, X.: Unified theory of acceptance and use of technology: a synthesis and the road ahead. J. Assoc. Inf. Syst. **17**(5), 328–376 (2016). https://doi.org/10.17705/1jais.00428

41. Davis, F.D.: Perceived usefulness, perceived ease of use, and user acceptance of information technology. MIS Q. 319–340 (1989). https://doi.org/10.2307/249008

42. Aguiar-Castillo, L., Clavijo-Rodriguez, A., Saa-Perez, D., Perez-Jimenez, R.: Gamification as an approach to promote tourist recycling behavior. Sustainability **11**(8), 2201 (2019). https://doi.org/10.3390/su11082201

43. Osborne, S.P.: From public service-dominant logic to public service logic: are public service organizations capable of co-production and value co-creation? Public Manage. Rev. **20**(2), 225–231 (2018). https://doi.org/10.1080/14719037.2017.1350461

44. Isaac, O., Abdullah, Z., Aldholay, A.H., Ameen, A.A.: Antecedents and outcomes of internet usage within organisations in Yemen: an extension of the Unified Theory of Acceptance and Use of Technology (UTAUT) model. Asia Pac. Manage. Rev. **24**(4), 335–354 (2019). https://doi.org/10.1016/j.apmrv.2018.12.003

45. Ajzen, I.: The theory of planned behavior. Organ. Behav. Hum. Decis. Process. **50**(2), 179–211 (1991)

46. Alalwan, A.A.: Mobile food ordering apps: an empirical study of the factors affecting customer e-satisfaction and continued intention to reuse. Int. J. Inf. Manage. **50**, 28–44 (2020). https://doi.org/10.1016/j.ijinfomgt.2019.04.008

47. Chuang, L.M., Chen, P.C., Chen, Y.Y.: The determinant factors of travelers' choices for pro-environment behavioral intention-integration theory of planned behavior, unified theory of acceptance, and use of technology 2 and sustainability values. Sustainability **10**(6), 1869 (2018). https://doi.org/10.3390/su10061869

48. Kang, J., Liu, C., Kim, S.H.: Environmentally sustainable textile and apparel consumption: the role of consumer knowledge, perceived consumer effectiveness and perceived personal relevance. Int. J. Consum. Stud. **37**(4), 442–452 (2013). https://doi.org/10.1111/ijcs.12013

49. Wu, J.J., Chen, Y.H., Chung, Y.S.: Trust factors influencing virtual community members: a study of transaction communities. J. Bus. Res. **63**(9–10), 1025–1032 (2010). https://doi.org/10.1016/j.jbusres.2009.03.022

50. Zhang, B., Du, Z., Wang, B., Wang, Z.: Motivation and challenges for e-commerce in e-waste recycling under "Big data" context: a perspective from household willingness in China. Technol. Forecast. Soc. Chang. **144**, 436–444 (2019). https://doi.org/10.1016/j.techfore.2018.03.001

51. Juaneda-Ayensa, E., Mosquera, A., Sierra Murillo, Y.: Omnichannel customer behavior: key drivers of technology acceptance and use and their effects on purchase intention. Front. Psychol. **7**, 1117 (2016). https://doi.org/10.3389/fpsyg.2016.01117

52. Holbert, R.L., Kwak, N., Shah, D.V.: Environmental concern, patterns of television viewing, and pro-environmental behaviors: integrating models of media consumption and effects. J. Broadcast. Electron. Media **47**(2), 177–196 (2003). https://doi.org/10.1207/s15506878job em4702_2

53. Huang, H.: Media use, environmental beliefs, self-efficacy, and pro-environmental behavior. J. Bus. Res. **69**(6), 2206–2212 (2016). https://doi.org/10.1016/j.jbusres.2015.12.031

54. Chen, S.Y.: True sustainable development of green technology: the influencers and risked moderation of sustainable motivational behavior. Sustain. Dev. **27**(1), 69–83 (2019). https://doi.org/10.1002/sd.1863

55. Brislin, R.W.: Back-translation for cross-cultural research. J. Cross Cult. Psychol. **1**(3), 185–216 (1970). https://doi.org/10.1177/135910457000100301

56. Hair Jr., J.F., Hult, G.T.M., Ringle, C., Sarstedt, M.: A Primer on Partial Least Squares Structural Equation Modeling (PLS-SEM). Sage Publications, Sage Publications (2017)

57. Khan, F., Ahmed, W., Najmi, A.: Understanding consumers' behavior intentions towards dealing with the plastic waste: perspective of a developing country. Resour. Conserv. Recycl. **142**, 49–58 (2019). https://doi.org/10.1016/j.resconrec.2018.11.020

58. Nunnally, J.C., Bernstein, I.H.: Psychometric Theory. McGraw Hill, New York (1994)

59. Bagozzi, R.P., Yi, Y.: On the evaluation of structural equation models. J. Acad. Mark. Sci. **16**(1), 74–94 (1988). https://doi.org/10.1007/BF02723327

60. Fornell, C., Larcker, D.F.: Evaluating structural equation models with unobservable variables and measurement error. J. Mark. Res. **18**(1), 39–50 (1981). https://doi.org/10.1177/002224378101800104

61. Hair, J.F., Anderson, R.E., Tatham, R.L., Black, W.C.: Análisis Multivariante, vol. 491. Prentice Hall, Madrid (1999)

62. Henseler, J., Ringle, C.M., Sarstedt, M.: A new criterion for assessing discriminant validity in variance-based structural equation modeling. J. Acad. Mark. Sci. **43**(1), 115–135 (2014). https://doi.org/10.1007/s11747-014-0403-8

63. Hair, J.F., Ringle, C.M., Sarstedt, M.: PLS-SEM: indeed a silver bullet. J. Market. Theory Pract. **19**(2), 139–152 (2011). https://doi.org/10.2753/MTP1069-6679190202

64. Shmueli, G., Ray, S., Estrada, J.M.V., Chatla, S.B.: The elephant in the room: predictive performance of PLS models. J. Bus. Res. **69**(10), 4552–4564 (2016). https://doi.org/10.1016/j.jbusres.2016.03.049

65. Payne, J.W.: Contingent decision behavior. Psychol. Bull. **92**(2), 382 (1982). https://doi.org/10.1037/0033-2909.92.2.382

66. Juaneda-Ayensa, E., Clavel San Emeterio, M., González-Menorca, C.: Person-organization commitment: bonds of internal consumer in the context of non-profit organizations. Front. Psychol. **8**, 1227 (2017). https://doi.org/10.3389/fpsyg.2017.01227

67. El Marrakchi, M., Bensaid, H., Bellafkih, M.: E-reputation prediction model in online social networks. Int. J. Intell. Syst. Appl. **9**(11), 17 (2017). https://doi.org/10.5815/ijisa.2017.11.03

68. Stern, P.C.: New environmental theories: toward a coherent theory of environmentally significant behavior. J. Soc. Issues **56**(3), 407–424 (2000). https://doi.org/10.1111/0022-4537.00175

69. Alalwan, A.A., Dwivedi, Y.K., Rana, N.P.: Factors influencing adoption of mobile banking by Jordanian bank customers: extending UTAUT2 with trust. Int. J. Inf. Manage. **37**(3), 99–110 (2017). https://doi.org/10.1016/j.ijinfomgt.2017.01.002

70. Khalilzadeh, J., Ozturk, A.B., Bilgihan, A.: Security-related factors in extended UTAUT model for NFC based mobile payment in the restaurant industry. Comput. Hum. Behav. **70**, 460–474 (2017). https://doi.org/10.1016/j.chb.2017.01.001

71. Verkijika, S.F.: Factors influencing the adoption of mobile commerce applications in Cameroon. Telematics Inform. **35**(6), 1665–1674 (2018). https://doi.org/10.1016/j.tele.2018.04.012

72. Baabdullah, A.M., Alalwan, A.A., Rana, N.P., Patil, P., Dwivedi, Y.K.: An integrated model for m-banking adoption in Saudi Arabia. J. Bank Market. **37**(2), 452–478 (2019). https://doi.org/10.1108/IJBM-07-2018-0183

73. Chen, M.F., Tung, P.J.: The moderating effect of perceived lack of facilities on consumers' recycling intentions. Environ. Behav. **42**(6), 824–844 (2010). https://doi.org/10.1177/0013916509352833

74. Lu, J., Yao, J.E., Yu, C.S.: Personal innovativeness, social influences and adoption of wireless Internet services via mobile technology. J. Strateg. Inf. Syst. **14**(3), 245–268 (2005). https://doi.org/10.1016/j.jsis.2005.07.003

75. Hofstede, G., Hofstede, G.J., Minkov, M.: Cultures and Organizations: Software of the Mind: Intercultural Cooperation and Its Importance for Survival. McGraw-Hill, New York (2010)

76. Bankole, F.O., Bankole, O.O., Brown, I.: Mobile banking adoption in Nigeria. Electron. J. Inf. Syst. Dev. Countries **47**(1), 1–23 (2011). https://doi.org/10.1002/j.1681-4835.2011.tb00330.x

77. Chiu, C.M., Wang, E.T.: Understanding Web-based learning continuance intention: the role of subjective task value. Inf. Manage. **45**(3), 194–201 (2008). https://doi.org/10.1016/j.im.2008.02.003

78. Sun, Y., Liu, L., Peng, X., Dong, Y., Barnes, S.J.: Understanding Chinese users' continuance intention toward online social networks: an integrative theoretical model. Electron. Markets **24**(1), 57–66 (2013). https://doi.org/10.1007/s12525-013-0131-9

79. Chopdar, P.K., Korfiatis, N., Sivakumar, V.J., Lytras, M.D.: Mobile shopping apps adoption and perceived risks: a cross-country perspective utilizing the Unified Theory of Acceptance and Use of Technology. Comput. Hum. Behav. **86**, 109–128 (2018). https://doi.org/10.1016/j.chb.2018.04.017

80. Riffai, M.M.M.A., Grant, K., Edgar, D.: Big TAM in Oman: exploring the promise of online banking, its adoption by customers and the challenges of banking in Oman. Int. J. Inf. Manage. **32**(3), 239–250 (2012). https://doi.org/10.1016/j.ijinfomgt.2011.11.007

81. Sharma, S.K., Al-Badi, A., Rana, N.P., Al-Azizi, L.: Mobile applications in government services (mG-App) from user's perspectives: a predictive modelling approach. Gov. Inf. Q. **35**(4), 557–568 (2018). https://doi.org/10.1016/j.giq.2018.07.002
82. Chong, A.Y.L.: Predicting m-commerce adoption determinants: a neural network approach. Expert Syst. Appl. **40**(2), 523–530 (2013). https://doi.org/10.1016/j.eswa.2012.07.068
83. Teo, A.C., Tan, G.W.H., Ooi, K.B., Hew, T.S., Yew, K.T.: The effects of convenience and speed in m-payment. Ind. Manage. Data Syst. **115**(2), 311–331 (2015). https://doi.org/10.1108/IMDS-08-2014-0231
84. Wong, C.H., Tan, G.W.H., Loke, S.P., Ooi, K.B.: Mobile TV: a new form of entertainment? Ind. Manage. Data Syst. **114**(7), 1050–1067 (2014). https://doi.org/10.1108/IMDS-05-2014-0146
85. Palos-Sanchez, P.R., Correia, M.B., Saura, J.R.: An empirical examination of adoption of mobile applications in Spain and Portugal, based in UTAUT. Int. J. Mobile Commun. **17**(5), 579–603 (2019). https://doi.org/10.1504/IJMC.2019.102085

# Developing Blockchain-Based Crowdfunding Model for Property Investment

Mohammed Ali Berawi[1,2]([⊠]) [iD], Mohamad Khaerun Zuhry Radjilun[2] [iD],
and Mustika Sari[2] [iD]

[1] Civil Engineering Department, Faculty of Engineering, Universitas Indonesia,
Depok 16424, Indonesia
`maberawi@eng.ui.ac.id`
[2] Center for Sustainable Infrastructure Development, Faculty of Engineering, Universitas
Indonesia, Kampus UI, 16424 Depok, Indonesia

**Abstract.** Crowdfunding is an approach that has been widely used to raise capital from individuals to finance a venture, known for its ability to reach a large pool of new capital gathered from the crowd. Though it has been extensively applied to finance property developments in real estate industry in some countries, crowdfunding method still has some drawbacks in regard to the accountability and transparency of the platform. Hence, this study attempts to propose a crowdfunding model for property investment that takes into account blockchain technology addressing those shortcomings. Both qualitative and quantitative methods are used to achieve the objectives of this study through literature review, questionnaire surveys, benchmarking study, and expert interviews. As a result, the features of crowdfunding platform utilizing blockchain technology that can improve the accountability and transparency are determined, and the design of the operational workflow for the blockchain-based property crowdfunding platform is proposed as an alternative to the conventional crowdfunding platform, therefore addressing financial issues in property investment.

**Keywords:** Blockchain · Crowdfunding · Property investment · Platform features · Operational workflow

## 1 Introduction

Property investment in Indonesia is mainly seen as profitable by investors as it was estimated to have an investible commercial property market size of US$226.23bn and a listed property market size of US$29.38bn in 2019 [14]. Even though that the demand for residential property is considered as steady [5], according to a study conducted by Rumah.com Affordability Sentiment Index [43], the people's purchasing power is still low due to some reasons such as high property prices, rising property prices, high down payment requirement for property purchase, and difficulties in getting housing loans from the bank.

© Springer Nature Switzerland AG 2021
D. Rodionov et al. (Eds.): SPBPU IDE 2020, CCIS 1445, pp. 23–39, 2021.
https://doi.org/10.1007/978-3-030-84845-3_2

Several emerging technology-based financial private companies provide choices of property financing for the public to address these issues, such as crowdfunding. Crowdfunding is a disruptive digital technology innovation widely known as an approach to raising capital from individuals to finance a venture initiated by a specific party through an intermediary internet-based platform [1, 37]. In this past decade, the crowdfunding funding scheme began to be widely used to finance infrastructure and built environment project developments in some countries [18, 28]. Owing to its flexibility and adaptability to the needs of different project characteristics [52], the crowdfunding approach characterized by low financing cost and high efficiency has also been implemented in the real estate industry for the past few years [25]. Crowdfunding in the real estate industry has brought advantages to property developers by giving them access to a large pool of new capital gathered from the crowd, who lend their money to project developers or take equity stakes as joint venture partners [24, 51].

Despite its abundant benefits, crowdfunding has its shortcomings, some of which are administrative and accounting challenges [47], weaker investor protection [6], the potential for fraud [16, 50], and transaction costs for the financial intermediaries [12, 38]. Completing the project on time may also be the challenge of post-crowdfunding due to the large numbers of investors involved as shareholders, as it is evidenced that over 85% of large crowdfunding projects are delayed later than expected [29].

Several studies have shown that solutions to most of the problems in traditional crowdfunding, which lack accountability and transparency, can be offered by blockchain technology, a distributed ledger that records transactions in a way that is secure, transparent, decentralized, cost-effective, and time-efficient [10, 40, 45]. Previous studies have also explored the implementation of blockchain technology in crowdfunding, such as the study conducted by Zhu & Zhou [53], which explored the practical implementation of blockchain to address trust issue in equity crowdfunding, while Saadat et al. [44] proposed Ethereum smart contracts to the crowdfunding site to provide more transparent transactions. Hassija et al. [19] developed a blockchain-based crowdfunding platform called Bitfund to reduce transaction costs and increase time efficiency issues by eliminating intermediaries.

Previous studies mentioned above typically only focused on addressing particular crowdfunding issues. Therefore, this paper attempts to further explore crowdfunding in property investment by developing a blockchain-based crowdfunding platform for residential property investment by taking into account the features of a crowdfunding platform that can address issues around accountability and transparency; hence property sales can be boosted.

## 2 Literature Study

### 2.1 Residential Property Investment

Housing is one of the basic human needs that has become one of the indicators for the people's standard of living and the needs for food and clothing, in which the prerequisite for comfortable, affordable, maintainable, and environmentally compliant housing should be met [20]. Furthermore, housing and its provision both in the short and long term have also become crucial to anticipate the increased human population growth caused by rapid urbanization [13].

In Indonesia, investment in residential properties keeps increasing every year since it is seen as a promising venture that stands out from other investment types. Low-interest rates, ever-increasing land prices, and significant market demand are the dominant factors that help property investment grow [46]. Moreover, through the updated Investment Law, the government policy bringing the principle of openness and accountability through supportive facilities and investment procedures has created a friendly investment atmosphere in Indonesia that fosters the growth in property investment [9].

The average national population growth rate of about 1.4% per year also significantly impacts the increase of residential property demands. However, the residential property sector development still cannot keep pace with population growth; therefore, there is still an issue on the imbalance of the housing demand and the number of housing provisions that need to be addressed [34]. Besides, it is still difficult for low-income people to buy the residential property. Therefore, some property financing schemes have been developed to help them in making residential house transactions, including housing loan schemes (KPR) provided by financial institutions, gradual cash installment schemes, and hard cash payment schemes. The latest development of property financing schemes that are quite popular recently is property crowdfunding, usually known as real estate crowdfunding. This scheme that lets property buyers lend money collected from individuals who benefit from the installment rates in return [30].

## 2.2 Crowdfunding

As an alternative approach to raising capital funds for a venture or a cause from individuals through an intermediary platform [1, 15], crowdfunding has been considered an imperative potency in entrepreneurial investment to foster economic empowerment and democratic advancement financial sector [3]. There are several types of crowdfunding with the return as its basis [11]. Investment-based crowdfunding or equity-based crowdfunding has financial returns as its objective [33], while debt-based crowdfunding receives interest with the repayment as a return [31]. On the other hand, crowdfunding utilized as a mechanism to raise social funds to finance non-profit endeavors such as social initiatives is called donation-based crowdfunding, and funds raised to finance community-building goals is called reward-based crowdfunding [32]. These crowdfunding models have different value creation in terms of economic and social values, as shown in Fig. 1 below.

|  |  | Economic Value | |
|---|---|---|---|
|  |  | **Low** | **High** |
| Social | **Low** | Reward | Equity |
| Value | **High** | Donation | Debt |

**Fig. 1.** Crowdfunding and value creation [26]

The initial concept of crowdfunding started from the principle of microfinance and crowdsourcing [41]. Over time, crowdfunding is now facilitated and supported by internet-based website services [27]; hence it became an alternative method to connect parties who need capital with investors whose funds through the internet platform as the intermediary [50]. Crowdfunding has been proven to be effective in funding, validating products, and developing significant value for the community, as many organizations that utilized crowdfunding as their financing scheme to raise capital have generated billions of income and generated more job opportunities [27].

The financial technology industry (Fintech) is now in the middle of a period of remarkable growth; the crowdfunding sector, in particular, has increased from 6.5% to 93.5% [35]. It has become popular in property investment because investors do not need to have a large amount of money to be able to invest in one property [17].

There were hundreds of equity-based crowdfunding platforms in Indonesia, several of which focused on financing property investment, such as KoinWorks, Investree, Amartha, Propertianda.com, Tavest, and Napro.id. Even though these crowdfunding platforms have many commons regarding financing types and investment choices, they have different features built into the platform's system. There are ten features identified from the crowdfunding platforms: protection fund, lenders insurance, default insurance, mentoring and monitoring for lenders, loan agreements, property lease guarantee, credit scoring, lenders' information, detailed financial report, and return on capital. The feature comparison for the property crowdfunding platforms in Indonesia can be seen in Table 1.

**Table 1.** Crowdfunding and the value creation (source: authors)

| Feature | Koin-Works | Investree | Amartha | Properti-anda.com | Tavest | Napro.id |
|---|---|---|---|---|---|---|
| Protection Fund | √ | | | | | |
| Lenders Insurance | | √ | | | | |
| Default Insurance | | | √ | | | |
| Mentoring and Monitoring for Lenders | | | √ | | | |
| Loan Agreements | | | | √ | | |
| Property Lease Guarantee | | | | | | √ |
| Credit Scoring | √ | √ | √ | | | |
| Lenders' Information | √ | √ | √ | √ | √ | √ |
| Detailed Financial Report | | | | | √ | |
| Return on Capital | | | | | √ | |

Equity-based crowdfunding platforms such as Propertianda.com, Tavest, and Napro.id have a more extended investment period than debt-based crowdfunding platforms such as Koinworks and Investree because equity-based crowdfunding has the advantage from leasing or selling funded properties such as the development of apartments or houses that can be leased later. Meanwhile, in debt-based crowdfunding such as Koinworks or Investree, the loaned investment funds for the financed properties must be returned with interest [17].

## 2.3  Blockchain

Blockchain is a digital innovation firstly introduced in 2008 as the underlying technology for cryptocurrency known as Bitcoin. Blockchain kept being developed by adding more functionalities as it is believed to have the potential to create new foundations for the economic and social systems [21], it kept being developed by adding more functionalities. Just a decade after its first introduction, it has now been adopted in various sectors and expanded from just being the digital platform of cryptocurrency to many more applications beyond currency, finance, and markets [48].

Blockchain is a distributed ledger that is used to store a growing list of records where each record is stored in a block. A new block will be created when there is a new record added to the network. Distributed ledger means that the database is strode scattered in various nodes that are updated simultaneously. In other words, blockchain is a group of people sharing data without an intermediary, but trust still can be built because all parties involved can see all the transactions that occur in the blockchain network.

Rules called consensus are needed in the blockchain network to prevent chaos in this distributed environment; thus, it should be followed by every transaction that occurs in the blockchain network. When new data is added to the network, a rule mechanism will check whether the transaction is valid or not. New transactions added to the network are linked to previous transactions using cryptography that cannot be altered after it is recorded, making blockchain resilient and as well as transparent at the same time [2].

The distributed nature of blockchain is a combination that matches the concept of crowdfunding, in which the needs in the crowdfunding approach can be met. Pledge-camp platform is an example of blockchain technology implemented in crowdfunding, in which it addresses the issues of accountability and transparency implementing Smart Contract [40]. It is a digital contract that can automatically self-execute when the predetermined requirements are met, aiming to fulfill payment terms, liens, confidentiality, and law. Smart contract can improve, automatize, and make a wide range of processes more efficacious [39], as it minimizes the existence of unintentional errors, reduces intermediaries, reduces contract fraud, administrative costs, and transaction costs in other contracts [49]. In crowdfunding, it can be used for data storage, contracting, and payment systems [36].

Several underlying blockchain platforms can be selected to build a blockchain project; some of them are paid, but some are open source, such as Ethereum, Hyperledger, MultiChain, and R3 Corda, enabling anyone to build decentralized applications not owned or controlled by a single entity but are powered by peers who run the nodes [4]. However, each of those platforms has its technical features critical for its adoption; hence selecting a good platform is a crucial step in creating a blockchain project [23].

One of the most popular cryptocurrency programming platforms in the smart contract ecosystem is Ethereum, a peer-to-peer network of virtual machines implemented to run distributed applications using its decentralized public blockchain to store, protect, and execute the contracts [22].

## 3 Research Methodology

To achieve its two objectives, the research methodology of this study is divided into two stages. The first stage was conducted to identify the features of the property crowdfunding platform utilizing blockchain technology that can increase its accountability and transparency. While the second stage was performed to develop the blockchain-based property crowdfunding model's operational workflow. This study combines quantitative and qualitative research methods through literature study and questionnaire surveys in the first stage and a benchmarking study to existing crowdfunding platforms followed by an expert interview to strengthen the research output in the second stage. The proposed research workflow is shown in Fig. 2.

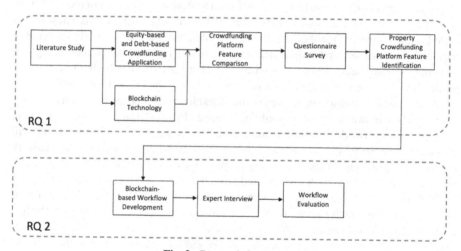

**Fig. 2.** Research framework

A comprehensive study of journals, books, conference papers, and reports from financial institutions and fintech companies regarding crowdfunding and blockchain was performed to obtain the research variables by comparing and identifying the accountability and transparency increasing features. The initial variables obtained were then written into a structured questionnaire used to collect data in the survey.

The questionnaire was spread out online to the targeted respondents. The non-probability sampling design with probability sampling statistical methods was the sampling used in this study. The respondents should fulfill one of these criteria, which include:

1. Has experience in or currently using crowdfunding platforms.
2. Has interest make a property investment.
3. Understands the concept of crowdfunding or work in the field of crowdfunding.

The result of the questionnaire survey was then used as the base for the workflow development of the blockchain-based crowdfunding model for the property investment transaction.

The questionnaire survey is divided into two main parts; the first one is identifying features that can increase the crowdfunding platform's accountability, while the latter is identifying features that can increase the transparency of the crowdfunding platform. The respondents were asked to select the most relevant features presented in the crowdfunding platform to increase both aspects of accountability and transparency.

The questionnaire survey results were used as the basis for the development of the workflow of blockchain implementation in the crowdfunding platform. It was then followed by interviews with experts and practitioners from the field of blockchain technology and residential property to validate the initially developed workflow for the blockchain-based crowdfunding model and obtain some valuable inputs for the improved final solution. The experts interviewed have prerequisite work experience in their respective fields of at least ten years and at least five years of experience for practitioners working with blockchain technology.

## 4 Results and Discussion

### 4.1 Background of the Respondents

After distributing the online survey questionnaire created on Google Form to mailing lists and social media platforms, a total of 361 questionnaires was completed. However, after checking all the filled survey questionnaires, only results from 238 filled questionnaires whose respondents met the criteria are considered valid for further analysis in data processing using the Statistical Package for Social Sciences (SPSS) 25.

From the total 238 respondents that completely fulfilled the questionnaire, 46.2% work in the finance industry, followed by 18.9% of them work in the public sector, 7.56% come from academic institutions, 7.15% in construction, 4.2% work in IT, and the rest 15.9% work in other industry.

### 4.2 Identification of Crowdfunding Platform Features

Ten variables regarding the features related to crowdfunding platform accountability and transparency were identified after the desk study was conducted. The identified variables were sent to survey respondents, the summary of which shown in Table 2.

**Statistical Analysis**
The 238 completed questionnaires were analyzed using statistical tests to examine their homogeneity, validity, and reliability.

The homogeneity test is carried out to determine each respondent's understanding regarding accountability and transparency feature selection based on the field of work.

**Table 2.** Features of Crowdfunding Platform

| Category | Variable | Feature | Description |
|---|---|---|---|
| Accountability | X1 | Protection Fund | Protection Fund feature on the KoinWorks platform serves as an insurance fund to return investors' money if borrowers failed to pay back. The amount of money returned depends on the credit scoring of the lender |
| | X2 | Default Insurance | The insurance feature provided by the Amartha platform to investors is by paying 1.5% of the funding so that if a default occurs, 75% of the investor's money will be returned |
| | X3 | Mentoring and Monitoring for Lenders | Mentoring and monitoring feature presented by the Amartha platform for lenders offers training regarding finances and provides supervision directly to the lenders |
| | X4 | Property Lease Guarantee | This feature is presented in Napro.id platform for investors to provide rental guarantees for the property invested |
| | X5 | Return on Capital | Return on capital is a feature provided by the Tavest platform to investors if the funds are not fully collected on the property, all the money invested will be returned to the investors |
| | X6 | Backers Protection | The Pledgecamp platform's insurance feature is provided for investors if the invested project fails to comply with the plan, the investment funds will be returned depending on the lender's agreement |
| Transparency | X7 | Information of Lenders | This feature is provided by almost all debt-based crowdfunding platforms to help investors assess potential lenders by showing lenders' information on campaigns |
| | X8 | Detailed Financial Report | The feature provided by the Tavest platform is used to report the use of funds on invested projects |
| | X9 | Credit Scoring | Almost all debt-based crowdfunding platforms provide this feature to help investors assess potential lenders by showing lenders' information on campaigns |
| | X10 | Campaign Deposits | The feature provided by the Pledgecamp platform requires money to be deposited by the lenders to support lenders' transparency. The lenders will be given a deposit if all their profile information has been provided |

Kruskal-Wallis test was selected since the profile of respondents has more than two categories. In testing the homogeneity, hypotheses determined in this test were:

$H_0$ = There is no difference in the perception of respondents with various work fields.

$H_1$ = There is a difference in the perception of respondents with various work fields.

Where $H_0$ is accepted if the value of Asym. Sig > level of significance ($\alpha = 0.05$), while $H_1$ is rejected if Asym. Sig < level of significance ($\alpha = 0.05$). The result of homogeneity tests obtained was Asym. Sig > 0.05 for X3, X4, X5, X6, X8, X9, and X10, meaning that there is no difference in the perception of respondents for these variables with various work fields, while result obtained for X1, X2, and X7 was Asym. Sig < 0.05, indicating that there is a difference in the perception of respondents for these variables with various work fields.

The validity test was computed with the requirement if $r_{count} > r_{table}$ and the sig. (2-tailed) < 0.05 in order for the instrument to be considered as valid. With 238 total respondents, degree of freedom (df) = N-2 = 238–2 = 236, and significance level of 5%, thus the $r_{table}$ is 0.1272. Based on the validity testing conducted using statistical analysis software, the $r_{count}$ value of all tested variables is higher than $r_{table}$ 0.1272, and the sig. (2-tailed) of all tested items are lower than the significance level 0.05. It can be concluded that all of the variables used as research instruments in this study are valid and can be used to collect the research data since all items fulfilled the decision-making basics of validity test.

Cronbach's alpha method was used to test the reliability with the following requirements:

a.  Cronbach's Alpha value ≤ 0.6 indicates that the questionnaire was not reliable
b.  Cronbach's Alpha value ≥ 0.6 indicates that the questionnaire is reliable

**Table 3.** Cronbach's alpha for accountability features

| Cronbach's Alpha | | | |
|---|---|---|---|
| 0.782 | | | |
| **Item-Total Statistics** | | | |
| Variabel | Scale Mean if Item Deleted | Scale Variance if Item Deleted | Corrected Item-Total Correlation | Cronbach's Alpha if Item Deleted |
| X01 | 2.6891 | 3.051 | 0.566 | 0.74 |
| X02 | 2.6849 | 3.052 | 0.566 | 0.74 |
| X03 | 2.7269 | 3.001 | 0.593 | 0.733 |
| X04 | 2.7689 | 3.157 | 0.488 | 0.759 |
| X05 | 2.7269 | 3.035 | 0.57 | 0.739 |
| X06 | 2.7059 | 3.306 | 0.4 | 0.78 |

**Table 4.** Cronbach's alpha for transparency features

| Cronbach's Alpha | | | | |
|---|---|---|---|---|
| 0.6 | | | | |
| **Item-Total Statistics** | | | | |
| Variabel | Scale Mean if Item Deleted | Scale Variance if Item Deleted | Corrected Item-Total Correlation | Cronbach's Alpha if Item Deleted |
| X07 | 1.7437 | 1.094 | 0.42 | 0.498 |
| X08 | 1.7521 | 1.082 | 0.433 | 0.488 |
| X09 | 1.6849 | 1.179 | 0.343 | 0.557 |
| X10 | 1.7143 | 1.184 | 0.328 | 0.569 |

Table 3 showed that the Cronbach's Alpha value of 0.782 for variables X1–X6, while Table 4 presented the Cronbach's Alpha value of 0.6 for variables X7–X10, indicating that the research instruments are reliable.

**Data Analysis**

The respondents need to clearly understand and well-informed regarding crowdfunding and residential property so that the answers given to the questionnaire survey are reliable, based on the respondents' knowledge. Therefore, the respondents were asked to fill their understanding level in the first part of the questionnaire using Likert scale. The results showed that the respondents' understanding of crowdfunding and residential property has an index of over 80% (see Table 5), indicating that they clearly understand.

**Table 5.** Index of respondents' understanding

| Component | Index | AVS (1) | D (2) | N (3) | A (4) | AS (5) | Index Total |
|---|---|---|---|---|---|---|---|
| Crowdfunding | Respondent | 0 | 0 | 5 | 29 | 204 | 97% |
| | Index | 0% | 0% | 1% | 10% | 86% | |
| Property | Respondent | 0 | 0 | 25 | 87 | 126 | 88% |
| | Index | 0% | 0% | 6% | 29% | 53% | |

**Legend:**

AVS = Agree Very Strongly     Index=0%-19.99% (AVS)
D = Disagree                  Index = 20%-39.99% (D)
N = Neutral                   Index = 40%-59.99% (N)
A = Agree                     Index = 60%-79.99% (A)
AS = Agree Strongly           Index = 80%-100% (AS)

The identification of features increasing the accountability and transparency in crowdfunding platform is described in the following figures. Figure 3 shows the dominant features for the accountability based on the respondents' selection: Protection Fund, Default Insurance, and Backers Protection, followed by Mentoring and Monitoring for

Lenders, Return on Capital, and Property Lease Guarantee. While Fig. 4 indicates that the most selected features for the transparency comprise Credit Scoring and Campaign Deposits, followed by Information of Lenders and Detailed Financial Report.

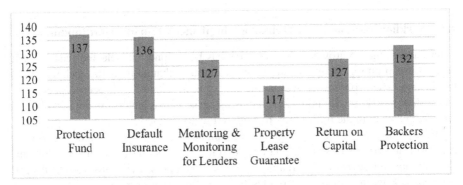

**Fig. 3.** Features increasing the accountability of crowdfunding platform

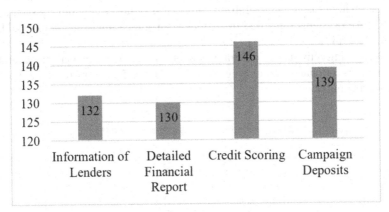

**Fig. 4.** Features increasing the transparency of crowdfunding platform

Based on the questionnaire survey results combined with literature and benchmark study, the authors chose the features of accountability and transparency on the crowdfunding platform to be further developed with the implementation of blockchain technology. These features are Protection Fund (X1), Mentoring and Monitoring for Lenders (X3), Return on Capital (X5), Information of Lenders (X7), Detailed Financial Report (X8), and Credit Scoring (X9).

Default Insurance and Backers Protection features have the same function as the Protection Fund, despite having a technically different method. Based on that fact, the authors decided to only choose one of those features with the highest number of respondents' selection, Protection Fund. Campaign Deposits also has the same function as the feature of Information of Lenders; however, many crowdfunding platforms use Information of Lenders features in carrying out these functions by requiring lenders to upload

34      M. A. Berawi et al.

their information before campaigning. Meanwhile, instead of requiring lenders' information, the Campaign Deposits feature requires depositing money before performing the campaign, in which the money will be returned once the lenders' information has been uploaded.

### 4.3 Workflow of Blockchain Technology Implementation in Crowdfunding

After identifying the crowdfunding features has been done, the selected features will be using to propose the model of the crowdfunding that utilizes the technology of blockchain. The proposed crowdfunding model is developed with peer-to-peer lending or debt-based crowdfunding scheme as its basis; hence the profits obtained by the investors is given in the form of interest offered by the lenders.

Ethereum, a full-featured programming language with its blockchain network, is used as the underlying blockchain platform for this crowdfunding platform development. The rationale for choosing Ethereum as the blockchain platform in this study is that, in contrast to Bitcoin, Ethereum provides features reprogrammed to facilitate complex logic. These features consist of the decentralized autonomous organization (DAO) that eliminates the third party [7], decentralized file storage (DFS) system that maintains files in a system without needing centralized silos that may not be completely secure [54], and ERC token systems supported by various open-source implementations and broad developer base [42].

Crowdfunding activities are generally classified into three main phases: the initiation phase, fundraising phase, and closing phase [8]. Figure 5 demonstrates the workflow diagram of the initiation phase.

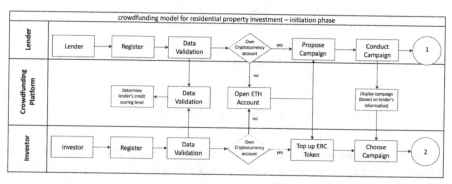

**Fig. 5.** Workflow diagram of the crowdfunding model's initiation phase

This phase begins with both lenders and investors register themselves in the platform, giving the required data such as general information, ID cards, photos with ID cards, fixed income letters, debt-free notes, financial reports for two years, and Taxpayer Identification Number (TIN) for individual lenders. For company/institutional lenders, the requirements comprise the company owner's ID card, company's establishment deed, trading business license, company's taxation identification number, two-year annual report. For investors, registration requirements are in the form of general information, an ID card, and a photo of themselves holding the ID card. All this data will be uploaded to the Ethereum network with DFS service then validated by the platform. After all of the data are verified, the platform will assess the credit scoring level obtained from the processing of the lender's information data using AI to determine the lender's risk level.

Next, a cryptocurrency wallet (account) used in the transactions on the crowdfunding platform will be created. If the lender already has an account, a fundraising campaign to buy property can be proposed to be conducted. On the other hand, if the investor already has an account, they can immediately top up the ERC tokens. During the campaign period, the lender's data is displayed on the platform so that investors can see and choose which investment campaigns they prefer. The information displayed in the campaign is the lender's profile, target funds, loan period, payment timeline, interest rate, credit scoring level, details of the project, or property to be funded.

Furthermore, the fundraising phase starts from fund collection until the lender receives the funds on to the second stage. Investors can start the funding using the ERC-20 token they own during the campaign period. Consequently, they will be then given rights by the platform through the Decentralized Autonomous Organization service to conduct discussions or voting on lenders regarding changes in the target time, target funds, payment timelines, interest rates will be obtained by investors etc. If the funds are collected according to the determined target time, the platform will transfer the funds to institutional lenders (companies) immediately to fund their residential projects. However, for individual lenders, the crowdfunding platform's collected funds will be transferred to the developer/seller of the residential property that the lender will buy. Documents from the property that have been paid for will then be used as collateral to the investors. The workflow of the fundraising phase is illustrated in Fig. 6 below.

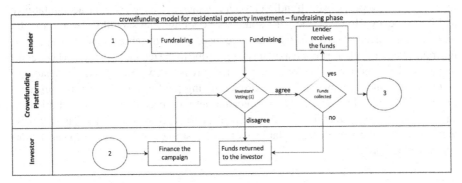

**Fig. 6.** Workflow diagram of the crowdfunding model's fundraising phase

The closing phase is the last stage that starts from funding the lender's property to the campaign's end. After the collected funds are transferred to the lender, the platform will conduct mentoring and monitoring activities to the lenders, aimed to observe the progress of using the funds and report and validate the progress on the platform to minimize the risk of fraud. Progress reports will then be uploaded to the Ethereum DFS network. After that, the lender's payment period to the investor begins. If the payment is not smooth, the investor will carry out a voting process to determine further steps for the said loan.

Moreover, when the borrower is considered the failure to pay, the investors will immediately receive protection funds or the insurance money according to the lender's credit scoring level. On the other hand, if the borrower can still pay according to the time until it is finished, the campaign is done. The workflow diagram of this phase can be seen in Fig. 7.

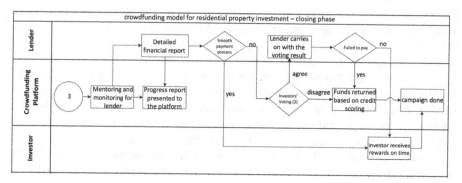

**Fig. 7.** Workflow diagram of the crowdfunding model's closing phase

Interviews were then conducted with three experts, which consist of academics in information technology, a blockchain infrastructure engineer, and a director of a real estate company based in Indonesia. This activity validated that the blockchain implemented on the proposed crowdfunding model can increase the platform's accountability and transparency. However, there are some shortcomings that should be overcome; the first one is that even though the costs can be reduced due to the absence of a third party in the transaction, the platform's costs can be higher due to the technology used. The second one is that DFS service from blockchain technology is a new system in the data storage service. The transparency level of the crowdfunding platform activities will surely be increased by uploading data on the blockchain network because everyone who is in it can validate the data, thereby increasing investor trust. However, as DFS is still new, there are still not many efficient platforms for data storage, resulting in slower upload and download processes.

## 5   Conclusion

Crowdfunding has gained popularity as an alternative approach to finance projects in the built environment industry by raising small amounts of capital from a large group of investors through internet-based technology due to its low-cost and highly efficient practice. However, it is still facing some challenges regarding the accountability and transparency of the method, in which Blockchain technology is proven to help address those issues through its decentralized and traceable organization. Through the questionnaire survey and benchmarking study, this research has identified the crowdfunding platform's features that can increase its accountability and transparency, which include protection fund, mentoring and monitoring for lenders, return on capital, and information of lenders detailed financial report, and credit scoring. These features were then applied in the proposed crowdfunding model for residential property investment that implements blockchain's Ethereum platform. The proposed crowdfunding model's workflow shows that implementing blockchain in crowdfunding has changed the way it works, offering the same services with a higher level of security, trust, transparency, and lower risk for all of the involved parties.

This study advises crowdfunding applications to consider adding features improving its accountability and transparency to increase the people's interests in using this approach as an alternative financing method in residential property investment. In this regard, this paper recommends the policymakers establish a policy instrument and implementation framework following the regulations for the future development of blockchain-based property crowdfunding platforms.

This study also encourages future studies to examine further the financial and institutional aspects of the proposed crowdfunding model to define the extent of its advantages to this crowdfunding financing scheme's practices for all stakeholders.

**Acknowledgement.** The authors would like to thank Indonesian Ministry of Research and Technology for the support given to this research.

# References

1. Ahlers, G.K.C., et al.: Signaling in equity crowdfunding. Entrep. Theory Pract. **39**(4), 955–980 (2015). https://doi.org/10.1111/etap.12157
2. Andoni, M., et al.: Blockchain technology in the energy sector: a systematic review of challenges and opportunities (2019). https://doi.org/10.1016/j.rser.2018.10.014
3. Assenova, V., et al.: The present and future of crowdfunding (2016). http://journals.sagepub.com/doi/10.1525/cmr.2016.58.2.125. https://doi.org/10.1525/cmr.2016.58.2.125
4. Bahga, A., Madisetti, V.K.: Blockchain platform for industrial Internet of Things. J. Softw. Eng. Appl. (2016). https://doi.org/10.4236/jsea.2016.910036
5. Bank Indonesia: Commercial Property Development Q2 2019. Jakarta, Indonesia (2019)
6. Bechter, C., et al.: From wisdom of the crowd to crowdfunding. J. Commun. Comput. **8**, 951–957 (2011)
7. Bracamonte, V., Okada, H.: An exploratory study on the influence of guidelines on crowdfunding projects in the ethereum blockchain platform. In: Ciampaglia, G.L., Mashhadi, A., Yasseri, T. (eds.) SocInfo. LNCS, vol. 10540, pp. 347–354. Springer, Cham (2017). https://doi.org/10.1007/978-3-319-67256-4_27
8. Branzov, T., Maneva, N.: Crowdfunding business models and their use in software product development. In: International Scientific Conference Informatics in Scientific Knowledge, Varna, Bulgaria (2014). https://doi.org/10.13140/RG.2.1.2738.8009.
9. Bunawan, P.: Foreign investment in Indonesia the legal aspects under the new indonesian investment law. Dialogia Iurid. J. Huk. Bisnis dan Investasi. (2017). https://doi.org/10.28932/di.v8i2.719
10. Cai, C.W.: Disruption of financial intermediation by FinTech: a review on crowdfunding and blockchain. Account. Financ. **58**(4), 965–992 (2018). https://doi.org/10.1111/acfi.12405

11. Cholakova, M., Clarysse, B.: Does the possibility to make equity investments in crowdfunding projects crowd out reward-based investments? Entrep. Theory Pract. **39**(1), 145–172 (2015). https://doi.org/10.1111/etap.12139
12. Darke, S.: To be or not to be a funding portal: why crowdfunding platforms will become broker-dealers. Hast. Bus. Law J. **10**(1), 183 (2014)
13. van Doorn, L., Arnold, A., Rapoport, E.: In the age of cities: the impact of urbanisation on house prices and affordability. In: Nijskens, R., Lohuis, M., Hilbers, P., Heeringa, W. (eds.) Hot Property, pp. 3–13. Springer, Cham (2019). https://doi.org/10.1007/978-3-030-11674-3_1
14. EPRA: Global Real Estate Total Markets Table Q4-2018. Brussels (2020)
15. Gajda, O., Walton, J.: Review of Crowdfunding for Development Initiatives, UK (2013). http://dx.doi.org/10.12774/eod_hd061.jul2013.gadja;walton
16. Gobble, M.A.M.: Everyone is a venture capitalist: the new age of crowdfunding (2012). https://www.jstor.org/stable/26586617. https://doi.org/10.5437/08956308X5504001
17. Hariyani, I., Yustisia, C.: Kajian Hukum Bisnis Jasa Crowdfunding Properti (Business law study of property crowdfunding services). J. Leg. Indones. **16**(1), 42–58 (2019)
18. Harrison, R.: Crowdfunding and the revitalisation of the early stage risk capital market: catalyst or chimera? (2013). https://www.tandfonline.com/doi/abs/10.1080/13691066.2013.852331. https://doi.org/10.1080/13691066.2013.852331
19. Hassija, V., et al.: BitFund: a blockchain-based crowd funding platform for future smart and connected nation. Sustain. Cities Soc. **60**, 102145 (2020). https://doi.org/10.1016/j.scs.2020.102145
20. Henilane, I.: Housing concept and analysis of housing classification. Balt. J. Real Estate Econ. Constr. Manage. **4**(1), 168–179 (2016). https://doi.org/10.1515/bjreecm-2016-0013
21. Iansiti, M., Lakhani, K.R.: The truth about blockchain (2017)
22. Jani, S.: An overview of ethereum & its comparison with bitcoin. Int. J. Sci. Eng. Res. **10**, 8 (2017)
23. Kuo, T.T., et al.: Comparison of blockchain platforms: a systematic review and healthcare examples (2019). https://doi.org/10.1093/jamia/ocy185
24. Leboeuf, G., Schwienbacher, A.: Crowdfunding as a new financing tool. In: Cumming, D., Hornuf, L. (eds.) The Economics of Crowdfunding, pp. 11–28. Springer, Cham (2018). https://doi.org/10.1007/978-3-319-66119-3_2
25. Maarbani, S.: Real estate crowdfunding: the digital evolution of real estate funding and investment. PricewaterhouseCoopers, Sydney (2015). https://doi.org/10.4018/978-1-7998-2448-0.ch031
26. Meyskens, M., Bird, L.: Crowdfunding and value creation. Entrep. Res. J. (2015). https://doi.org/10.1515/erj-2015-0007
27. Mollick, E.: Crowdfunding as a font of entrepreneurship: outcomes of reward-based crowdfunding. In: Cumming, D., Hornuf, L. (eds.) The Economics of Crowdfunding, pp. 133–150. Springer, Cham (2018). https://doi.org/10.1007/978-3-319-66119-3_7
28. Mollick, E.: The dynamics of crowdfunding: determinants of success and failure. SSRN Electron. J. (2012). https://doi.org/10.2139/ssrn.2088298
29. Mollick, E., Robb, A.: Democratizing innovation and capital access: the role of crowdfunding. Calif. Manage. Rev. **58**(2), 72–87 (2016). https://doi.org/10.1525/cmr.2016.58.2.72
30. Montgomery, N., et al.: Disruptive potential of real estate crowdfunding in the real estate project finance industry: a literature review (2018). https://doi.org/10.1108/PM-04-2018-0032
31. Moss, T.W., et al.: The effect of virtuous and entrepreneurial orientations on microfinance lending and repayment: a signaling theory perspective. Entrep. Theory Pract. **39**(1), 27–52 (2015). https://doi.org/10.1111/etap.12110
32. Motylska-Kuzma, A.: Cost of crowdfunding as a source of capital for the small company. In: 18th International Academic Conferences, August, pp. 461–473 (2015)

33. Motylska-Kuzma, A.: Equity-based crowdfunding in polish conditions. In: 4th International Multidisciplinary Scientific Conference on Social Sciences and Arts SGEM2017, MODERN SCIENCE. Stef92 Technology (2017). https://doi.org/10.5593/sgemsocial2017/13/s03.035
34. Napitupulu, S., et al.: Kajian Perlindungan Konsumen Sektor Jasa Keuangan: Perlindungan Konsumen Pada Fintech (Studies on consumer protection in the financial services sector: consumer protection at Fintech). Jakarta, Indonesia (2017)
35. Nugroho, A.Y., Rachmaniyah, F.: Fenomena Perkembangan Crowdfunding Di Indonesia. Ekonika J. Ekon. Univ. Kediri. **4**(1), 34–46 (2019). https://doi.org/10.30737/ekonika.v4i1.254
36. Oliva, G.A., Hassan, A.E., Jiang, Z.M.: An exploratory study of smart contracts in the Ethereum blockchain platform. Empirical Softw. Eng. **25**(3), 1864–1904 (2020). https://doi.org/10.1007/s10664-019-09796-5
37. Ordanini, A., et al.: Crowd-funding: transforming customers into investors through innovative service platforms. J. Serv. Manag. **22**(4), 443–470 (2011). https://doi.org/10.1108/095642311 11155079
38. Peck, M.E.: Blockchains: how they work and why they'll change the world. IEEE Spectr. **54**(10), 26–35 (2017). https://doi.org/10.1109/MSPEC.2017.8048836
39. Penzes, B.: Blockchain Technology in the Construction Industry: Digital Transformation for High Productivity, London (2018). https://doi.org/10.13140/RG.2.2.14164.45443
40. Pledgecamp: Pledgecamp. The next generation of crowdfunding (2019)
41. Poetz, M.K.M.S.: The value of crowdsourcing: can users really compete with professionals in generating new product ideas? J. Prod. Innov. 245–256 (2012). http://dx.doi.org/10.12774/eod_hd061.jul2013.gadja;walton
42. Roth, J., et al.: The tokenization of assets: using blockchains for equity crowdfunding. SSRN Electron. J. (2019). https://doi.org/10.2139/ssrn.3443382
43. Rumah.com: Rumah.com Property Affordability Sentiment Index H2 2019. Jakarta (2019)
44. Saadat, M.N. et al.: Blockchain based crowdfunding systems. Indones. J. Electr. Eng. Comput. Sci. **15**(1), 409–413 (2019). https://doi.org/10.11591/ijeecs.v15.i1.pp409-413
45. Schatsky, D., Muraskin, C.: Beyond Bitcoin. Blockchain is coming to disrupt your industry. https://www2.deloitte.com/us/en/insights/focus/signals-for-strategists/trends-blockchain-bitcoin-security-transparency.html. Accessed 27 July 2020
46. Sean, S.L., Hong, T.T.: Factors affecting the purchase decision of investors in the residential property market in Malaysia. J. Surv. Constr. Prop. (2014). https://doi.org/10.22452/jscp.vol 5no2.4
47. Sigar, K.: Fret no more: Inapplicability of crowdfunding concerns in the internet age and the jobs act's safeguards (2012). http://www.sec.gov/spotlight/dodd
48. Swan, M.: Blockchain: Blueprint for a new economy (2015). https://doi.org/10.1017/CBO 9781107415324.004
49. Tapscott, D., Tapscott, A.: Blockchain Revolution: How the Technology Behind Bitcoin is Changing Money, Business, and the World (2018). https://doi.org/10.1080/10686967.2018.1404373
50. Valančienė, L., Jegelevičiūtė, S.: Valuation of crowdfunding: benefits and drawbacks. Econ. Manag. **18**(1), 39–48 (2013). https://doi.org/10.5755/j01.em.18.1.3713
51. Vogel, J.H., Moll, B.S.: Crowdfunding for real estate. Real Estate Financ. J. 5–16 (2014)
52. Zhang, T., Chen, J.: The feasibility study of introducing crowdfunding into the PPP model of profit-making enterprises. Presented at the July 1 (2017). https://doi.org/10.2991/essaeme-17.2017.240
53. Zhu, H., Zhou, Z.Z.: Analysis and outlook of applications of blockchain technology to equity crowdfunding in China. Financ. Innov. **2**(1), 1–11 (2016). https://doi.org/10.1186/s40854-016-0044-7
54. Zichichi, M., et al.: On the Efficiency of Decentralized File Storage for Personal Information Management Systems (2020)

# Ultra High Speed Broadband Internet and Firm Creation in Germany

Kirill Sarachuk$^{(\boxtimes)}$ ⓘ, Magdalena Missler-Behr ⓘ, and Adrian Hellebrand ⓘ

Brandenburg University of Technology, Erich-Weinert-Str. 1, 03046 Cottbus, Germany
kirill.sarachuk@b-tu.de

**Abstract.** Latest changes in broadband internet provision, as it was expected, will improve the life quality of millions of people around the world and will simplify the business routine, so that even new companies would be able to compete with well-established ventures. However, while there are numerous examples how modern digital infrastructure increases productivity or economic growth, very few papers try to link the availability of high-speed internet and the appearance of new firms. With our previous research at the municipality level in Brandenburg, we started to cover this research gap for the case of Germany. Our current study is focusing on possible effects of (ultra-)broadband provision on firm entry rates in 401 German administrative districts and independent cities. Results of our regression analysis demonstrate significance of basic broadband availability for the whole country (and very strong for less-urbanized administrative districts), while the ultra-high-speed connections have a weak and negative influence on new business formations. Hence, the advanced digital infrastructure does not seem like an important prerequisite for a better entrepreneurial milieu, at least in the case of Germany.

**Keywords:** Broadband · Internet · ICT · Entrepreneurship · Germany

## 1 Introduction

More than ten years ago Europe has made its first step in a digital transformation by introducing Digital Agenda for Europe [1], a comprehensive long-term policy aimed at creating a network-based economic area with a vibrant broadband infrastructure in its core. With its main postulate – to promote competitiveness and social inclusion across Europe, – the policy stipulated that every household and enterprise should be able to benefit from next-generation access networks (NGAN) at least at speed 30 Mbps by the end of year 2020. It was expected that, on one hand, it will foster potential entrepreneurs to establish more business ventures as even small companies would be able to find their place in the sun. On the other hand, Digital Agenda had been considered to improve the life standards of millions of citizens thanks to augmented social and healthcare digitalized services, to the possibility to make a better choice between various brands or even to the simplified data exchange between governments and households (e-Government).

Despite the decision taken by European Commission was in the zeitgeist, already at the early stages many experts claimed that the proposed targets seem to be very ambitious.

© Springer Nature Switzerland AG 2021
D. Rodionov et al. (Eds.): SPBPU IDE 2020, CCIS 1445, pp. 40–56, 2021.
https://doi.org/10.1007/978-3-030-84845-3_3

First, the lack of information and communication technology (ICT) skills among the potential users was obvious, given that more than a half of the European citizens had no or just basic internet skills [2]. Even by 2020 most of the EU population had insufficient digital competence: nearly sixty percent had at least basic digital skills [3], while at the same time more and more jobs require them; therefore, just a modest number of users may benefit from augmented ICT infrastructure to a full extend. Second, the speed of digital development is unevenly distributed across the EU: while some countries as Malta and Netherlands reported the accomplishment, Italy, Romania and Slovakia were far behind even with the first milestone [4, 5]. The major reason for that is the general unpreparedness of institutions for radical changes [6] resulting in very slow place in taking important decisions and absence of clear development policies.

Although better digital technologies, of course, are beneficial for the society in terms of productivity gains [7] or higher employment [8], it does not necessarily mean that advanced broadband infrastructure entails excessive firm entries or higher economic performance, as the rule *the faster, the better* does not perform well in this context [9]. One of the main reasons is exactly that better internet speed usually transforms into economic gains along with the presence of high-skilled employees [10] – which is utterly possible whether people have only elementary digital competence. Although, there is a well-known bias that advanced ICTs usually turn into higher formation rates as far as cooperation between companies becomes much easier and their ways how to reach customers – much shorter. Still, better digital infrastructure cannot guarantee neither better entrepreneurial milieu nor significant advantages for potential entrants, not least due to increased market competition [11].

Despite an increased interest towards the problem how better ICT infrastructure reshapes the business patterns globally, scholars still cannot answer confidently whether a high-speed internet may foster potential founders to set up new businesses [12]: the up-to-date results are rather contradictive, depending on the area of study. We are mostly interested in the case of Germany, which, on the one hand, has a good total economic performance and above EU-average digital infrastructure, but also drastic differences between different parts of the country (mainly former western and eastern states). In our previous paper [13], we found rather a negative impact of ultra-broadband internet on firm-birth rate for peripheral municipalities in Brandenburg and no statistical significance for other entities in the vicinity to the capital city Berlin. This paper is to continue our research for the case of Germany by taking into consideration the whole country, but now at a more aggregated level (administrative districts and autonomous cities), with larger and better data available. This is not the first study trying to look at the mentioned problem from the regional level [14, 15], but will be the first that tried to highlight not only the differences between West and East Germany as many traditional studies, but rather western, eastern, and southern (the richest) parts of the country.

The paper unfolds as follows. We give a short overview on the existing scientific literature in Sect. 2. In the following Sect. 3 we describe in the nutshell the case of Germany: the broadband availability and status of entrepreneurship. Section 4 deals with the methodology and datasets we used for our research. Results of our regression analysis with a discussion part follow in Sect. 7. Finally, Sect. 8 concludes the paper with a brief summary.

## 2  Literature Background

Over the past decades newer ICTs – prior to all broadband and ultra-broadband internet – have transformed the global business routine dramatically as far as they changed the economical meaning of national borders and distances [16], reduced the costs of communication and made the information exchange much easier than before [17]. Because of that, even not only established ventures but also smaller companies were able to expand quickly and compete on the same market alongside market leaders [18]. Furthermore, the recent COVID-19 pandemic clearly demonstrated that the demand for ultra-high-speed digital infrastructure will grow further, for instance, due to the multiple advantages of teleworking both for companies and employees [19].

A positive effect of telecommunication infrastructure on economies and business activities was already observed in 1970s and 1980s [20, 21] and later on was also confirmed for dial-up era [22] and broadband connections [7]; hence, high-speed internet connections are regarded as the important driver of prosperity [23] and overall economic growth [24–28]. Precisely, ICTs are an important prerequisite for productivity increase [29, 30], however it applies only in case when high-skilled employees [10, 31] or investments into digital competence and strategic transformation [32] are present. At the same time, some researchers were not able to find a significant enabling effect of modern ICTs [33] or argue that only moderate broadband coverage levels, not the gigabit networks that are expected to be deployed by year 2025 in some European countries, will be finally profitable for society [34].

Still, the better broadband infrastructure does not only have a spurring effect on growth and productivity, but also may boost R&D and innovation activities [35] or infrastructural and technological improvements inside companies [36]. Along with that, ICTs may contribute to the overall economic gains through higher employment rates [37].

At the same the role of telecommunication infrastructure on entrepreneurship is not investigated quite rigorously so far [12]. Very few scholars tried to explain the role of ICTs in setting up new business, as for instance McCoy et al. [38] revealed higher statistical significance of DSL and fiber broadband connections for setting up a venture in high-tech business sectors or Hasbi [39] who observed higher firm entry rates in French municipalities with ultra-high-speed internet provision.

What is, however, a bit better known to the date is the possible presence of spatial variety effects of high-speed internet availability onto firm birth rates [40]. Some scholars mention that modern ICTs can serve as enabling powers for peripheral areas [41] and rural regions [42], especially if such areas are benefiting from a closer location to the metropolitan areas or have higher population amounts – it is argued that internet serves as a substitute to handshake interaction in such a case. However, some studies draw an opposite picture stating that any positive effects could be observed for core regions only [43] or there are minor differences with respect to the impact of ICTs on firm location [44]. With respect to the case of Germany, there is an obvious research gap; previous studies are rather outdated and refer to the older communication technologies as DSL [45]. In our recent research [13] for Brandenburg, the region surrounding the capital state Berlin and, as could be expected, benefiting from its successful location, we found either negative impact of ICTs on firm birth rate for peripheral areas or no statistical significance for other administrative entities.

# 3  The Case of Germany

We already mentioned that one of the prime objectives of the Digital Agenda was to promote entrepreneurship across Europe. It was proposed that a better digital infrastructure could not only serve as a medium assisting small and medium companies at their early stages (for instance, thanks to newer business models) and encouraging their cooperation with well-established business entities in the later phases, but also as an enabling power for innovation activity and smart networks enhancing productivity and economic growth. The successful realization of those plans, however, would be possible in case when the development speed is relatively equal to avoid possible disparities between different regions. Obviously, that is not true for the entire Europe, but for Germany we may state that even after more than thirty years after the re-unification many pronounced differences may be observed between former western and eastern German parts.

Former regions of German Democratic Republic (excepting Berlin) are falling behind by various criteria, prior to all in terms of wealth of households and entrepreneurship. While East Germany's industrial milieu is mainly described as low-tech with a modest share of high-skilled human capital [46], West Germany has a stronger focus on labor- and knowledge intensive industries with higher rates of networking and a growing demand on qualified labor force – which has a direct impact on the salaries and wealth of the inhabitants. Given a modest number of research facilities, it can be clearly seen why the eastern part of the country lacks R&D clusters and generally demonstrates weak regional competitiveness.

Still, while the differences between East and West Germany are well-studied in the scientific literature, rather a growing distinction between north and the south becomes a more important topic in the last years. The major reason is that the technological and knowledge-intensive companies tend to locate themselves in the two richest southern regions of Germany – Baden-Württemberg and Bayern – not least due to the higher number of institutions that contribute to the creation of research clusters and, consequently, better entrepreneurial milieu. Even other former western regions are also advanced in terms of research and development or innovative activities, exactly the southern part of Germany benefits from the best living standards and physical infrastructure, but also collaboration between companies.

## 3.1  Broadband Availability in Germany

Germany was among the first countries that successfully integrated the goals set by Digital Agenda for Europe in national broadband development plans already in 2009. Since that time, the country was steadily working on the availability of its digital infrastructure and demonstrated rather a good performance in internet acceptance due to better digital competence in the population [47]. However, that does not hold true for digitalization of government as just a very small share of services is currently available online for all German citizens.

Despite the recent developments, the overall quality of broadband internet leaves a lot to be desired. Previously, Deutsche Telekom dominated in the German (A)DSL market with a share nearly by ninety percent [48] which turned out into a problem that the majority of German households still have this type of connection with a standard rate not exceeding 30 Mbit/s, but what is more disgusting, at a very high price. As of 2016, less that ten percent of users had a possibility to enjoy the ultra-high-speed internet via fiber-optic, but finally just nearly 1% of all users preferred this type of connection [49], and by 2020 the amount of such households increased up to nearly five percent [50]. Hence, it is worth to say that the high-speed internet in Germany is not fast, but also quite expensive.

To mention is also that the speed of development of NGAN is not equal throughout the whole country: bigger cities and highly urbanized regions usually enjoy better digital infrastructure as they can possibly attract new firms. However, much more differences may be found between West and East German states: Fig. 1 clearly demonstrates that even the availability of basic broadband (at rates of 30 Mbit/s or faster) is slightly worse in the eastern part (and generally is better in the southern states), but much more pronounced differences could be observed for ultra-broadband internet connections (200 Mbit/s or faster). Thus, it is quite clear that even prime objectives set by Digital Agenda for Europe were not accomplished. It is still possible that Germany is able to meet the second goal and provide the ultra-broadband connections to at least half of households as it was planned by the end of 2020, but it is clear that this goal may be achieved before all for former western states, while the eastern part will remain far behind.

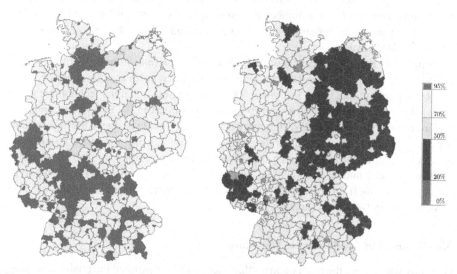

**Fig. 1.** Broadband avaliability in Germany at 30 Mbps (left) and 200 Mbps (right) in 2018. Source: Own compilation based on Broadband Atlas (Breitbandatlas) for Germany

## 3.2 Entrepreneurial Milieu in Germany

Historically regarded as a major driver of the European economy, Germany, showing much better economic performance that the EU average, surprisingly does not have the best entrepreneurial milieu due to several reasons. First, the levels of self-employment remain very modest, especially in the eastern part of the country where a partial inheritance of the socialistic attitude towards business can still be observed. As the result, loss of business activity becomes a general trend for East Germany in the recent years with the number of exits exceeding the number of new business formations (see Appendix 1). Contrast to that, very few areas in West Germany experience such a negative firm dynamics. In addition, north-western, and south-eastern parts are showing better performance (see Fig. 2), not the least due to the higher level of cooperation between business and research entities, but also regional competitiveness that pursues entrepreneurs to invest more in R&D activities [51]. Still, it is fair to note that East Germany has much less urbanized areas and generally less inhabitants that also reflect the business formation rates.

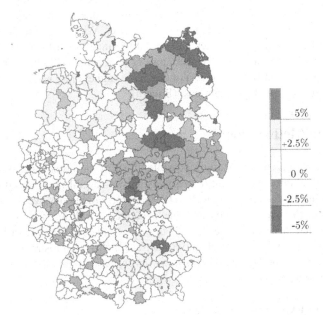

**Fig. 2.** Changes in firm population in Germany in 2018. Source: Own compilation based on Regional Database for Germany

Another reason explaining the low entrepreneurial spirit among German potential founders is the low entrepreneurial culture [52]. Even though several nationwide projects promoting business culture and entrepreneurship were launched in the past decade – and some of them were successful, – this problem still requires much more attention. A possible way to promote the business spirit is the introduction of entrepreneurial education at the universities, which is obviously lacking in the entire country, but before all in the eastern part of the country where even the number of research institutions is much lower compared to the country average.

One more serious problem the potential founders encounter has more psychological roots: the social perception of entrepreneurs and business culture is very mixed in the German society. In fact, just successful businesses deserve much attention from the public, while possible failures or market exits are often connected with reputational losses [53]. This is rather common to the eastern business philosophy and extremely differs from the American way of doing business where even multiple failures do not necessarily prevent a potential founder to make another try (so called *no pain, no gain* principle). Due to that reason, many founders in Germany first gather necessary skills and knowledge over a long period before setting up an own firm; an average successful entrepreneur is reflected in the mindset as a fifty-year-old or more, with many years of work experience [54]. That is slightly different from the United States where youth is encouraged to open the firm earlier (for instance, tech ventures are founded on average five years after graduation) and an average nascent entrepreneur is nearly forty-two years old [55].

Currently, a prevalent number of German firms are micro-enterprises with less than ten employees (nearly ninety percent of the total firm population), although their share is slightly shrinking against small and medium enterprises. With respect to the firm entries and exits, the firm birth rate was generally positive since the last economic recession in the end of 2000s, but, again, the eastern part of the country is falling far behind and losing more business entities. In the rest of Germany, higher entry rates are observed for north-west regions as well for south-east (Bayern).

## 4   Data and Methodology

The aim of our research is to study a possible effect of the broadband availability on entrepreneurial activity for the entire Germany. Given several pronounced differences in high-speed internet coverage, as well as business formation rates, we run a regression analysis first for the whole country and then separately for different parts of Germany (for more detailed description see Sect. 5) in order to look at possible difference. For the purpose of our study, we developed a dataset for 401 administrative districts and independent cities (*Landkreise und kreisfreie Städte*) based on the latest available data from multiple sources.

### Dependent and Explaining Variables
The dependent variable, *Firm Entries*, was retrieved from the Registry of Companies for Germany. It shows the total number of entries for all economic sectors, irrespectively of their size or type of formation, for every administrative entity. The data for this variable were taken for 2018, the last year available in the whole dataset.

Our explaining variables, *Basic Broadband* and *Ultra Broadband*, represent the ultra-broadband coverage ratio for all the observations at rates 30 Mbit/s and 200 Mbit/s respectively. We decided not to exclude any of those due to some evidence in the literature that higher speeds do not necessarily contribute for better entrepreneurial milieu. Both variables were retrieved manually from the Broadband Atlas for Germany (Breitbandatlas Deutschland) available at the website of the Federal Ministry of Transport and Digital Infrastructure.

### Population Data
Not only digital infrastructure, but rather wide set of factors contributes for firm formation rates. Before all, the data about population and its structure are often mentioned in the scientific literature to study the effects of broadband on economic patterns, productivity,

or growth. For instance, Armington and Acs [56] mentioned a positive effect of economically active population and density rates on firm formation. Clearly, higher population amounts raise more potential entrepreneurs, however the higher population density may turn into tougher competition that can hamper entrepreneurial activity. Additionally, the existing literature stresses the phenomenon of necessity entrepreneurship, a situation when people without job have no other opportunity but to try to set up an own business, so higher unemployment rated may contribute to higher number of entries. Finally, areas which are culturally diverse also possibly attract more potential entrepreneurs [57].

Based on the aforesaid, we included into our dataset the variables *Population* representing the economically active population aged 15 to 65, *Population Density*, *Unemployment* standing for the share of unemployed inhabitants in each administrative unit and *Migrants* for share of people with migratory background. All these parameters originate from the open-source Regional Database for Germany (Regionaldatenbank Deutschland) and retrieved for year 2018.

**Physical Infrastructure**
Another important driver for entrepreneurship is the availability of physical infrastructure, such as density of highways and railroads, or the distances to the nearest sea- and airports. As in case of Germany, a distance to airport may have much more importance as there are only sixteen air gates operating international flights, while a well-organized system of roads and railways is available for in-country transportation. Therefore, we included a manually calculated with the help of Google Maps variable *Distance to Airport*, and also retrieved a parameter *Transport Surface* representing the density of transportation infrastructure in square kilometers from the Regional Database for Germany (Regionaldatenbank Deutschland).

**Universities and Higher Education**
We already mentioned in Sect. 2 that a spurring effect of broadband is sometimes observed in case of presence of employees with a high level of competences. Research centers and universities, hence, seem to be very important for a better entrepreneurial environment in regions as they produce and disseminate basic knowledge that companies subsequently may use for own purposes, but also prepare a high-skilled workforce. In our dataset, we included a variable *Distance to University*, the distance in kilometers from the administrative district to the nearest of 295 state subsidized research units calculated manually with the help of Google Maps. Whether an administrative unit has at least one university, we assume that the distance is equal to zero.

Much more complicated is the situation with the information for high-skilled workforce itself, as the last census data gathered in Germany dates back to 2011 and, as we suggest, is pretty outdated for our analysis. As a way to overcome this problem, we decided to include the variable *People with Abitur* that shows the share of inhabitants in 2018 holding a German high school graduation certificate (or equivalent) which is an important requirement to continue the education at the university. Precisely, only those people who obtained such a certificate are allowed to pursue for a least bachelor's degree which, in accordance with the European Qualifications Framework, means that a holder is able to manage complex professional activities. However, we would rather interpret the result with caution as not every holder of the high school graduation certificate in Germany finally accomplishes the study at a university.

**GDP and Personal Income**
Contrast to our previous study for municipalities in Brandenburg, we included some financial data to our analysis. Higher amounts of GDP and available personal income are usually connected with a wealth of households [40], and whether households have more spare financial resources, the higher is the chance that they use these funds to set up an own firms. Therefore, we added both variables – *GDP per Capita* and *Personal Average Annual Income* – into our dataset. Both variables were retrieved from the open-source regional statistical databases for all German federal states.

**Estimation Equation**
As the dimensions of our variables differ significantly, we use a natural log transformation in order to normalize the data and make the outcome more plausible. To avoid a problem that some values in our data (distances to nearest *university* and *airport*) may be equal to zero, we add a constant value (1) to each value of those variables. Similar to our previous study and earlier works by Kolko [58] and Fabritz [45], we use a traditional OLS estimated regression method for the following equation:

$$\ln(Entries_{2018}) = \alpha + \beta_1 \times \ln(BB_{2018}) + \beta_2 \times \ln(UBB_{2018}) + \gamma \times \ln(Controls_{2018}) + \varepsilon,$$

where $Entries_{2018}$ is our dependent variable, $BB_{2018}$ and $UBB_{2018}$ stand for basic and ultra-broadband and $Controls_{2018}$ is the vector for population data, physical infrastructure, universities and financial results of regions and households. The description statistics for all variables is provided in Appendix 2.

# 5   Results and Discussion

Results of our analysis are presented in Table 1. The first, Sample A, covers all Germany. Samples B and C represent the former western and eastern part of the country, respectively. The two richest southern regions, Baden-Württemberg and Bayern, are represented by the Sample D and would be analyzed apart from western part of the country (not included into the Sample B). Finally, Sample E refers to independent cities and Sample F to administrative districts.

**Digital Infrastructure**
The relation between basic high-speed internet coverage and firm entries is positive and significant in all six models. The effect is, however, more pronounced in the south of Germany, and at the same time a bit weaker in the West Germany compared to the eastern part. As we mentioned in Sect. 3, the broadband infrastructure is better developed in the southern part of the country, the outcome does not seem to be much surprising. However, it may indicate that further improvements in high-speed internet availability have be done in West and East Germany so that other regions would be able to reach the same level of business activity as in Baden-Württemberg and Bayern.

In addition, the effect in independent cities is much higher than in more rural administrative districts, although the outcome is very strong significant just in the Sample F. This result could be also expected as bigger cities attract more people, have better infrastructure, and generally offer augmented possibilities for new entrants. Still, that does not necessarily mean that rural areas have to deserve much less attention in terms of the development of high-speed internet connection. Summing up, the basic broadband infrastructure seems to be necessary for new entrants in the entire Germany, but still contribute to more market entries in the richest regions (Baden-Württemberg and Bayern) and independent cities.

At the same time, the ultra-broadband connections appear to be insignificant in all models with negative coefficients. These results are quite similar to our previous study for Brandenburg (where we found a negative effect of ultra-high-speed internet for peripheral municipalities), rather that here we find no significance irrespective of the area. That again allows to prompt a conclusion that even need for speed and gigabyte era may be attractive for end-users and entrepreneurs, there is no essence in them to this point of time.

**Population**

Along with basic broadband, the economically active population is another factor that has a strong and positive significance in all models which is quite obvious and stays in line with previous findings [31, 38]. At the same time, the population density is significant and negatively associated (similar to the study of Fabritz [45]) with the firm entries in the general model as well as for South Germany and in case of districts. This outcome could be interpreted in the following way: the more inhabitants are living in a particular area, the higher is the number of potential entrepreneurs and, possibly, tougher is the competition, which finally may have a negative impact on firm entries.

With respect to the unemployment, a negative and strong correlation is observed for the general model and also South Germany, while a slightly negative but weak correlation was found for the administrative districts. While a positive correlation may indicate a greater labor availability (see for example papers by Coughlin [59] and McCoy et al. [38]), a negative result may indicate a presence of necessity entrepreneurship, or the situation when unemployed people set up a firm as they see no other opportunity (see the study of Hasbi [39] for French municipalities). While for the more rural districts the outcome is rather predictable, it is a bit stunning that we observed a strong importance for advanced and more technologically driven southern part where the entry barriers are much higher. However, as we could not separate the sample for tech and non-tech companies, we may suggest that the possible existence of necessity entrepreneurship in South Germany refers more to less technological driven firm formations.

Not too much papers tried to study the connection between the share of inhabitants with migratory background and firm creation (for instance, Hart and Acs [57] observed the importance of immigrants in founding of high-tech companies). Hence, it was quite interesting to observe a positive and significant relation of the share of migrants with respect to the firm entries in all models excepting West Germany, but the effect is more pronounced for the eastern part (compared to the south) and for administrative districts (compared to cities). Generally, migrants in Germany try to obtain a university degree and then tend to stay in more urbanized areas as they usually get more career possibilities.

Table 1. Results of regression analysis (OLS estimated, variables natural log transfromed)

| DV: *Entries* 2018 | (A) | (B) | (C) | (D) | (E) | (F) |
|---|---|---|---|---|---|---|
| $BB_{2018}$ [% coverage] | 0.447*** [0.120] | 0.331** [0.166] | 0.552** [0.234] | 0.806** [0.344] | 1.949** [0.888] | 0.507*** [0.112] |
| $UBB_{2018}$ [% coverage] | −0.0279 [0.0470] | −0.018 [0.0645] | −0.103 [0.116] | −0.137 [0.0843] | −0.212 [0.234] | −0.0521 [0.0454] |
| Population [thsd] | 1.152*** [0.0495] | 0.983*** [0.0854] | 1.143*** [0.117] | 1.273*** [0.0872] | 1.330*** [0.153] | 1.190*** [0.0504] |
| Density [thsd per sqkm] | −0.0812** [0.0722] | 0.0664 [0.0601] | −0.0679 [0.0799] | −0.152** [0.061] | −0.0977 [0.104] | −0.163*** [0.0401] |
| People with Abitur [% share] | 0.076 [0.117] | 0.119 [0.207] | 0.260 [0.282] | 0.113 [5.074] | −0.495* [0.265] | 0.298** [0.124] |
| Unemployment [% share] | −0.105*** [0.0297] | −0.0158 [0.0516] | −0.110 [0.110] | −0.168*** [0.0626] | −0.112 [0.0848] | −0.0543* [0.0318] |
| Migrants [% share] | 0.150*** [0.0189] | −0.0108 [0.0403] | 0.313*** [0.0849] | 0.141** [0.0612] | 0.102** [0.051] | 0.161*** [0.0196] |
| Transport Surface [sqkm] | −0.145*** [0.0497] | 0.0361 [0.0862] | 0.0162 [0.118] | −0.334*** [0.0855] | −0.361** [0.158] | −0.115** [0.0486] |
| Distance to Airport [km] | −0.011 [0.0118] | −0.0166 [0.0166] | 0.0122 [0.0247] | −0.0148 [0.0221] | −0.0227 [0.0185] | −0.0282* [0.0166] |
| Distance to University [km] | −0.0037 [0.00687] | 0.0134 [0.0104] | −0.0169 [0.0169] | −0.000858 [0.0111] | 0.0162 [0.0161] | −0.0053 [0.00721] |
| Avg. Annual Income [thsd. EUR] | 0.337*** [0.122] | 0.365** [0.164] | 0.421 [0.594] | 0.149 [0.185] | 0.0611 [0.231] | 0.727*** [0.146] |
| GDP per Capita [thsd. EUR] | 0.00267 [0.0344] | 0.0414 [0.0459] | 0.129 [0.164] | −0.0264 [0.0587] | 0.0330 [0.0718] | −0.00914 [0.0442] |
| Constant | −9.144*** [0.764] | −8.218*** [1.129] | −11.58*** [2.519] | −10.05*** [1.739] | −13.56*** [3.968] | −11.52*** [0.807] |
| Area | Whole | West | East | South | Cities | Districts |
| Observations R-squared Adj. R-squared | 401 0.953 0.952 | 184 0.959 0.956 | 77 0.966 0.960 | 140 0.952 0.947 | 107 0.971 0.968 | 297 0.948 0.945 |

Standard errors in brackets *p: 0.1 **p: 0,05 ***p: 0,01

However, we suggest that in case of entrepreneurship people with migratory background possibly have more chances in less-competitive rather than in high-competitive parts of the country.

Finally, we also did not find too much proof that a higher amount of people holding a German high school graduation certificate (or equivalent) contribute to the firm entries (with an exception for administrative districts). Generally, a greater amount of population with higher education is positively associated with higher firm entry rates [10, 39]. Still, as we mentioned, our variable *People with Abitur* does not necessarily show the amount of high-qualified people and, hence, should be interpreted with caution. A real number of university graduates could possibly give much more reliable information, but, unfortunately, there is no fresh data available to the point of time.

**Other Control Variables**
Regarding the physical infrastructure, it is interesting to mention that a positive but weakly significant relation for the distance to the nearest airport is observed for the Sample F, but not in the rest of the models. At the same time, the density of roads and railroads is strongly and negative significant in all models excepting Samples B and C. It may seem stunning at the first glance that smaller number of firms tend to enter areas with better physical infrastructure, but we may interpret that in a way due to higher density of transportation lines the area becomes more attractive for potential entrants (and also customer) which finally may initiate the tougher competition between firms. That also may explain why we have observed no significance with respect to the distance to the nearest university – while firms usually benefit from a closer location to the research unit, this is also a favorable environment for competition.

Finally, we did not find any significance for the variable *GDP per capita*, but at the same time we observed a positive and highly significant connection between firm formations and financial success of regions and households in the general model. This general outcome stays in line with the previous findings [40, 60]. However, among all areas the average annual income seems to be significant only for West Germany and not important in the rest of the country, and for administrative districts but not for the independent cities. This result may be explained in a way that many small and successful start-ups, so called *hidden champions*, that usually require a lot of financial resources at the earlier stages, are located in the western part. At the same time, financial resources do not play too much role in the southern part due to its overall higher wealth, and in case of East Germany possibly due to fact household generally have much less money to set up an own business. Still, while in the big cities the entry barriers are usually higher due to the tougher competition, higher wealth of households in rural administrative districts seem to play much more important role in a decision to set up an own business.

# 6 Conclusion

This paper continued our last research with a study of possible effects of broadband availability on firm birth rate, but now for the whole Germany. A major outcome from our regression analysis is that a high-speed internet provision generally contributes for more entries, but the effect is more pronounced in the southern part of the country, and much higher in independent cities than in administrative districts. We suggest that the focus for high-speed internet provision improvements should lay in West and East Germany, but also more in rural administrative districts, so they have a chance for catching-up in terms of firm entries, while the richer southern part and cities require something more than the speed of communications. Along with that, an ultra-broadband provision does not seem to be such important, so we have to bear in mind that high-speed connections do not guarantee their unconditional adoption or exploitation [41].

Several outcomes of our analysis (with respect to the population data or availability of physical infrastructure) stay in line with our last research [13]. We were able to use some extra information (financial performance of regions and households, however, did not deliver us too much important results) available at the administrative districts and independent cities level, but its scarcity (as for the number of high-skilled employees in the region) still remains the most important limitation to our research. Furthermore, this paper analyses the situation at one certain period of time and does not give any clue how the situation changed over time (eg. at the beginning of implementation of Digital Agenda for Europe and at its latest stage). Hence, our further study has to answer this question and draw a more comprehensive conclusion towards the problem whether a broadband internet has a relation to firm creation over a longer time period.

## Appendix 1. Number of Active Firms in German States, 2015–2018

Source: Own compilations based on Regionaldatenbank Deutschland [61].

| | 2018 | 2017 | 2016 | 2015 | ± since 2015 |
|---|---|---|---|---|---|
| Baden-Württemberg | 472 664 | 468 349 | 464 279 | 467 205 | ▲ 1,17% |
| Bayern | 633 624 | 622 156 | 619 311 | 618 906 | ▲ 2,38% |
| Bremen | 28 341 | 26 465 | 27 109 | 27 122 | ▲ 4,49% |
| Hamburg | 107 426 | 102 996 | 102 930 | 102 444 | ▲ 4,86% |
| Hessen | 275 043 | 274 577 | 273 161 | 272 617 | ▲ 0,89% |
| Niedersachsen | 294 576 | 287 936 | 288 515 | 287 180 | ▲ 2,58% |
| Nordrhein-Westfalen | 721 390 | 717 282 | 716 044 | 711 967 | ▲ 1,32% |
| Rheinland-Pfalz | 159 461 | 159 542 | 160 552 | 159 809 | ▼ 0,22% |
| Saarland | 37 018 | 36 798 | 37 536 | 37 159 | ▼ 0,38% |
| Schleswig-Holstein | 123 616 | 122 409 | 123 037 | 122 923 | ▲ 0,56% |
| Berlin | 187 981 | 182 214 | 179 663 | 175 180 | ▲ 7,31% |
| Brandenburg | 98 230 | 98 293 | 98 425 | 97 531 | ▲ 0,72% |
| Mecklenburg-Vorpommern | 60 820 | 62 081 | 61 184 | 63 223 | ▼ 3,80% |
| Sachsen | 163 252 | 165 174 | 166 387 | 166 447 | ▼ 1,92% |
| Sachsen-Anhalt | 71 389 | 73 831 | 75 457 | 76 024 | ▼ 6,10% |
| Thüringen | 79 594 | 81 757 | 82 603 | 83 302 | ▼ 4,45% |
| Total, of those: | 3 514 425 | 3 481 860 | 3 476 193 | 3 469 039 | ▲ 1,31% |
| micro-firms | 89,10% | 89,30% | 89,52% | 89,75% | |
| small and medium firm | 10,46% | 10,27% | 10,06% | 9,84% | |
| macro firms | 0,44% | 0,43% | 0,42% | 0,41% | |

## Appendix 2. Descriptive Statistics for Variables (Natural Log Transformed)

| Variable | Description | Mean | St. Dev. | Min | Max |
|---|---|---|---|---|---|
| Entries 2018 | Number of new firms in 2018 | 7.08769 | 0.74117 | 5.26786 | 10.63960 |
| $BB_{2018}$ | Coverage by basic broadband | 4.50684 | 0.09918 | 3.93183 | 4.60517 |
| $UBB_{2018}$ | Coverage by ultra-broadband | 4.16500 | 0.31811 | 2.63906 | 4.59512 |
| Population | Economically active population aged 15–65 years | 11.98103 | 0.66040 | 10.44203 | 15.10881 |
| Density | Population density thsd. per sqkm | 5.62688 | 1.10331 | 3.59800 | 8.46296 |
| People with Abitur | People holding a German high school graduation certificate (or equivalent) | 3.80824 | 0.07331 | 3.58629 | 4.07414 |
| Unemployment | Share of unemployed people | 1.44564 | 0.44904 | 0.33647 | 2.54945 |
| Migrants | Share of people with migratory background | 2.44411 | 0.67412 | 0.66096 | 3.78213 |
| Transport Surface | Density of transportation infrastructure in sqkm | 3.55168 | 0.80394 | 1.14422 | 5.15063 |
| Distance to Airport | Distance to the nearest airport in km | 3.88291 | 0.85908 | 0 | 5.28244 |
| Distance to University | Distance to the nearest university in km | 2.13196 | 1.55169 | 0 | 4.50103 |
| Avg Annual Income | Average annual net income per inhabitant | 3.10719 | 0.11201 | 2.79190 | 3.66423 |
| GDP per capita | GDP per capita in thsd | 3.52325 | 0.32369 | 2.79972 | 5.14810 |

## References

1. European Commission: A digital agenda for Europe. Publications Office of the European Union (2010)
2. European Commission: Digital Agenda Scoreboard 2011. Pillar 6: Digital Competence in the Digital Agenda (2011)
3. European Commission: Digital Economy and Society Index 2020. Human capital (2020)
4. IHS Markit: Broadband Coverage in Europe 2018: Mapping progress towards the coverage objectives of the Digital Agenda (2018)

5. Matteucci, N.: L'investimento nelle reti NGA a larga banda: la questione settentrionale. Economia e Politica Industriale (2014)
6. McAleese, M., Bladh, A., Bode, C., Muehlfeit, J., Berger, V., Petrin, T.: Report to the European Commission on new modes of learning and teaching in Higher Education. Publications Office of the European Union, 6–7 (2014)
7. Grimes, A., Ren, C., Stevens, P.: The need for speed: impacts of internet connectivity on firm productivity. J. Prod. Anal. **37**, 187–201 (2012)
8. Lehr, W., Gillett, S., Sirbu, A.: Measuring Broadband's Economic Impact, Broadband properties, December 2005 (2005)
9. Bai, Y.: The faster, the better? The impact of internet speed on employment. Inf. Econ. Policy **40**, 21–25 (2017)
10. Akerman, A., Gaarder, I., Mogstad, M.: The skill complementarity of broadband internet. Q. J. Econ. **130**, 1781–1824 (2015)
11. Kandilov, I., Renkow, M.: Infrastructure investment and rural economic development: an evaluation of USDA's broadband loan program. Growth Change **41**, 165–191 (2010)
12. Sarachuk, K., Missler-Behr, M.: ICT, economic effects and business patterns: a text-mining of existing literature. In: Proceedings of the 3rd International Conference on Computers in Management and Business, pp. 40–45 (2020)
13. Sarachuk, K., Missler-Behr, M.: Is ultra-broadband enough? The relationship between high-speed internet and entrepreneurship in Brandenburg. Int. J. Technol. **11**, 1103–1114 (2020)
14. Audretsch, D., Dohse, D., Niebuhr, A.: Cultural diversity and entrepreneurship: a regional analysis for Germany. Ann. Reg. Sci. **45**, 55–85 (2010)
15. Fabritz, N.: ICT as an Enabler of Innovation. Evidence from German Microdata. Ifo Working Paper (2015)
16. Thurik, R., Stam, E., Audretsch, D.: The rise of the entrepreneurial economy and the future of dynamic capitalism. Technovation **33**, 302–310 (2013)
17. Vu, K.: ICT as a source of economic growth in the information age: empirical evidence from the 1996–2005 period. Telecommun. Policy **35**, 357–372 (2011)
18. Jung, W.-J., Lee, S.-Y.T., Kim, H.-W.: Are information and communication technologies (ICTs) displacing workers? The relationship between ICT investment and employment. Inf. Dev. **36**, 520–534 (2020)
19. Martins, S.S., Wernick, C.: Regional differences in residential demand for very high bandwidth broadband internet in 2025. Telecommun. Policy **45**, 102043 (2021)
20. Greenstein, S., Spiller, P.: Modern telecommunications infrastructure and economic activity: an empirical investigation. Ind. Corp. Change **4**, 647–665 (1995)
21. Roller, L.-H., Waverman, L.: Telecommunications infrastructure and economic development: a simultaneous approach. Am. Econ. Rev. **91**, 909–923 (2001)
22. Hagén, H.-O., Glantz, J., Nilsson, M.: ICT use, broadband and productivity. In: Yearbook on Productivity'(Statistics Sweden), pp. 37–70 (2008)
23. Myovella, G., Karacuka, M., Haucap, J.: Digitalization and economic growth: a comparative analysis of Sub-Saharan Africa and OECD economies. Telecommun. Policy **44**, 101856 (2020)
24. Gruber, H., Hätönen, J., Koutroumpis, P.: Broadband access in the EU: an assessment of future economic benefits. Telecommun. Policy **38**, 1046–1058 (2014)
25. Koutroumpis, P.: The economic impact of broadband: evidence from OECD countries. Technol. Forecasting Soc. Change **148**, 119719 (2019)
26. Mayer, W., Madden, G., Wu, C.: Broadband and economic growth: a reassessment. Inf. Technol. Dev. **26**, 128–145 (2020)
27. Pradhan, R.P., Bele, S., Pandey, S.: Internet-growth nexus: evidence from cross-country panel data. Appl. Econ. Lett. **20**, 1511–1515 (2013)

28. Vu, K., Hanafizadeh, P., Bohlin, E.: ICT as a driver of economic growth: A survey of the literature and directions for future research. Telecommun. Policy **44**, 101922 (2020)
29. Cardona, M., Kretschmer, T., Strobel, T.: ICT and productivity: conclusions from the empirical literature. Inf. Econ. Policy **25**, 109–125 (2013)
30. Mačiulytė-Šniukienė, A., Gaile-Sarkane, E.: Impact of information and telecommunication technologies development on labour productivity. Procedia Soc. Behav. Sci. **110**, 1271–1282 (2014)
31. Mack, E., Faggian, A.: Productivity and broadband: the human factor. Int. Reg. Sci. Rev. **36**, 392–423 (2013)
32. Colombo, M., Croce, A., Grilli, L.: ICT services and small businesses' productivity gains: an analysis of the adoption of broadband Internet technology. Inf. Econ. Policy **25**, 171–189 (2013)
33. Haller, S., Lyons, S.: Broadband adoption and firm productivity: evidence from Irish manufacturing firms. Telecommun. Pol. **39**, 1–13 (2015)
34. Briglauer, W., Gugler, K.: Go for Gigabit? First evidence on economic benefits of high-speed broadband technologies in Europe. JCMS J. Common Market Stud. **57**, 1071–1090 (2019)
35. Xu, X., Watts, A., Reed, M.: Does access to internet promote innovation? A look at the US broadband industry. Growth Change **50**, 1423–1440 (2019)
36. Koutroumpis, P.: The economic impact of broadband on growth: a simultaneous approach. Telecommun. Policy **33**, 471–485 (2009)
37. Jorgenson, D., Motohashi, K.: Information technology and the Japanese economy. J. Jpn. Int. Econ. **19**, 460–481 (2005)
38. McCoy, D., Lyons, S., Morgenroth, E., Palcic, D., Allen, L.: The impact of broadband and other infrastructure on the location of new business establishments. J. Reg. Sci. **58**, 509–534 (2018)
39. Hasbi, M.: Impact of very high-speed broadband on company creation and entrepreneurship: empirical Evidence. Telecommun. Policy **44**, 101873 (2020)
40. Parajuli, J., Haynes, K.: spatial heterogeneity, broadband, and new firm formation. Qual. Innov. Prosperity **21**, 165–185 (2017)
41. Mack, E.: Businesses and the need for speed: the impact of broadband speed on business presence. Telematics Inform. **31**, 617–627 (2014)
42. Kim, Y., Orazem, P.: Broadband internet and new firm location decisions in rural areas. Am. J. Agric. Econ. aaw082 (2016)
43. Capasso, M., Cefis, E., Frenken, K.: Spatial differentiation in industrial dynamics. The case of the Netherlands (1994-2005). Tijdschrift voor economische en sociale geografie **107**, 316–330 (2016)
44. Mack, E., Grubesic, T.: Broadband provision and firm location in Ohio: an exploratory spatial analysis. Tijdschrift voor economische en sociale geografie **100**, 298–315 (2009)
45. Fabritz, N.: The impact of broadband on economic activity in rural areas: evidence from German municipalities. Ifo Working Paper (2013)
46. Ragnitz, J.: Explaining the East German productivity gap: the role of human capital. Kiel Working Paper (2007)
47. European Commission: Digital Economy and Society Index 2019. Human capital (2019)
48. Gries, C.-I.: The development of DSL markets in international context. WIK Diskussionsbeitrag (2004)
49. BMWi: Digital Strategy 2025 (2016)
50. Koptyug, E.: Households with access to fibre-optic communications in Germany 2007–2020. Statista (2020). https://www.statista.com/statistics/469139/fibre-optic-connections-households-with-access-germany/. Accessed 24 Mar 2021
51. Kosfeld, R., Mitze, T.: The role of R&D-intensive clusters for regional competitiveness. Joint Discussion Paper Series in Economics (2020)

52. Bittorf, M.: Germany's entrepreneurial culture: strengths and weaknesses. KfW Econ. Res. **39** (2013)
53. Wyrwich, M., Stuetzer, M., Sternberg, R.: Entrepreneurial role models, fear of failure, and institutional approval of entrepreneurship: a tale of two regions. Small Bus. Econ. **46**, 467–492 (2016)
54. Sternberg, R., von Bloh, J: Global Entrepreneurship Monitor (GEM). Country Report Germany 2016. Eschborn and Hanover: RKW and Institute of Economic and Cultural Geography … (2017)
55. Azoulay, P., Jones, B.F., Kim, J.D., Miranda, J.: Age and high-growth entrepreneurship. Am. Econ. Rev. Insights **2**, 65–82 (2020)
56. Armington, C., Acs, Z.: The determinants of regional variation in new firm formation. Reg. Stud. **36**, 33–45 (2002)
57. Hart, D., Acs, Z.: High-tech immigrant entrepreneurship in the United States. Econ. Dev. Q. **25**, 116–129 (2011)
58. Kolko, J.: Broadband and local growth. J. Urban Econ. **71**, 100–113 (2012)
59. Coughlin, C.C., Terza, J.V., Arromdee, V.: State characteristics and the location of foreign direct investment within the United States. Rev. Econ. Stat. 675–683 (1991)
60. Sutaria, V., Hicks, D.A.: New firm formation: dynamics and determinants. Ann. Reg. Sci. **38**, 241–262 (2004)
61. Regionaldatenbank Deutschland: Unternehmensregister-System (URS) (2018). https://www.regionalstatistik.de/genesis/online/data?operation=statistic&code=52111

# Industrial, Service and Agricultural Digitalisation

# Information Modeling Technology as the Integrating Basis of the Development Investment Process

Nadezhda Kvasha[1] (ID), Ekaterina Malevskaya-Malevich[2]([⊠]) (ID),
and Svetlana Kornilova[3] (ID)

[1] Bonch-Bruevich St. Petersburg State University of Telecommunications, St. Petersburg, Russia
[2] Peter the Great St. Petersburg Polytechnic University, St. Petersburg, Russia
malevskaia@spbstu.ru
[3] St. Petersburg State University of Aerospace Instrumentation, St. Petersburg, Russia

**Abstract.** The full life cycle of a development investment project is an organized set of stage-by-stage projects, the subject area of which is represented by a multitude of organizations of the construction complex. These entities are characterized by low affiliation, which, when it is necessary to ensure the fulfillment of independent tasks in the absence of additional mechanisms ensuring integral effectiveness of the project, can lead to ineffective solutions. The article proposes a thesis that integration of project subjects, understood as an interaction characterized by a higher degree of cooperation, ensures the efficiency of the entire development investment process due to the rational use of resources and increasing labor productivity. An actively developed digital basis for the integration of independent project entities based on the market mechanism of interaction is the information modeling technology of a development investment project, formed according to the principles of the BIM (Building Information Modeling) concept. Due to the presence of a broad range of externalities during the implementation of this approach, both at the microeconomic level of different stages of the project, and at the meso-level in the development of cities and territories, the article suggests that the level of implementation of information modeling technology will not be able to achieve efficient values without effective mechanisms of internalization, which, in turn, require scientific comprehension and elaboration. The purpose of the study is to substantiate the implementation and develop proposals for the advancement of information modeling technology as an integrating basis for the development investment process, ensuring its effectiveness. At the same time, the final problem solved to achieve this goal is to establish the approaches and results of the participation of the public side in the process of introducing and developing information modeling technology in the development investment complex.

**Keyword:** Information modeling technology · Investment project · IPD partnership

© Springer Nature Switzerland AG 2021
D. Rodionov et al. (Eds.): SPBPU IDE 2020, CCIS 1445, pp. 59–72, 2021.
https://doi.org/10.1007/978-3-030-84845-3_4

# 1  Introduction

The full life cycle of the development investment project is an organized set of stage-by-stage projects, such as the design of the construction facility, its creation, operation, liquidation (redevelopment), etc. In this consideration, the subject area of the project is represented by a multitude of organizations of the construction complex (construction and installation, repair and construction, design and survey; enterprises producing building materials, structures, products; enterprises producing and renting out construction and roadwork equipment and tools; service infrastructure organizations, etc.). As a rule, the specified set of entities is characterized by low affiliation, which, when it is necessary to ensure the fulfillment of independent tasks in the absence of additional mechanisms facilitating integral effectiveness of the project, can lead to ineffective solutions. Thus, interactions between entities in traditional contracting methods are characterized by adherence solely to their own goals within the project, lack of interest in the results of other entities, inability (including economic) to exchange resources, etc. [which has been noted, among others, in 1, 4]. It may be argued that the integration of the project's subjects (from the initial pre-investment stage to the redevelopment and/or liquidation of the construction facility), understood here as an interaction characterized by a higher degree of cooperation aimed at creating favorable economic conditions, ensures the efficiency of the entire development investment process through the rational use of resources and increased productivity, which, at the current low values of these indicators in the Russian Federation [5], is particularly relevant.

Preliminarily stating the priority of the market mechanism for coordinating the activities of economic agents, it should be noted that the necessary condition for its successful implementation is the near-zero value of transaction costs, which is ensured by the existence of a unified information space of the project, which in the current state of information communication development is digital in nature [5]. Actively developing digital basis of integration of independent design entities based on the market mechanism of interaction is the technology of information modeling of a development investment project. Considering the wide range of externalities during the implementation of this approach, both at the microeconomic level of the various stages of the project and at the meso-level in the development of cities and territories, it can be suggested that the level of implementation of information modelling technology will not be able to achieve efficient values without effective internalization mechanisms, which in turn require scientific understanding and development.

Thus, the purpose of the study is to justify the implementation and development of proposals for the advancement of information modeling technology as the integrating basis for the development investment process that ensures its effectiveness. In order to achieve this goal, the following problems were solved:

– theoretical and empirical testing of the hypothesis stating that there is a direct relationship between the level of integration of project participants and the efficiency of the development investment process;
– rationalization of the mechanism coordinating the activities of economic agents and of an effective model for the implementation of development investment projects based on the information modeling technology;

- identification of the formation specifics of the implementation effects of the information modeling technology in the development investment process and justification of the key role of the public party;
- establishment of the approaches and outcomes of the public party's contribution to the implementation and development of information modeling technology in the development investment complex.

## 2  Materials and Methods

### 2.1  Hypothesis Testing

The theoretical justifications for the proposed hypothesis are based on the results of studies [for example, 1–4, 18, 22, 25], which indicate that the processes of forming the results of development investment projects considered throughout their life cycle are characterized by high probability of losses from underinvestment under conditions of mutual effects ("internal" externalities), when the increase in cost and/or duration of one stage of the project affects the quality of the results of another. For example, the effect of increasing the duration and cost of design work is more evident during the construction phase, ensuring the quality of the facility being created, as well as reducing the duration and cost of construction work. The effect of the costs of ensuring the quality and durability of materials within the construction stage is mostly realized during the operation stage. Moreover, the investments of some entities (for example, operating organizations) can provide additional effects to others (for example, users of the constructed facility).

The empirical validation of the hypothesis was based on the results of existing practical studies in the field of analysis of the nature and degree of interactions of different parties throughout the life cycle of the project and their relationship with productivity and the transfer of innovations in the construction industry [in particular, 3, 8, 21]. One of the main obstacles to the innovative development of the development investment sector can be "classification of investments in innovation and their results during different in relation to the "owner" (executor) stages and operations of the development investment cycle". This conclusion is based on the fact that only a small percentage of investment in innovation provides feedback at the same stage of the development investment process. In addition, the greatest economic effect is created by investments in innovation, which are manifested in product and technological changes at all stages of the life cycle of a development project.

The study [8] analyzed the relationship of general contractors (customers) with other actors in the development investment process in five EU countries: Denmark, France, Germany, Sweden and the United Kingdom. This study presents qualitative characteristics of the level of integration for each pair "general contractor (customer) – another party" (weak, medium, strong, very strong). To convert the qualitative characteristics provided in [8] into quantitative indicators, we have calculated the final indices of the level of integration of the development investment process in the $i^{th}$ country ($J_{PCi}$). To calculate $J_{PCi}$, each $n^{th}$ line of interaction (interaction with a certain group of entities) was assigned specific indices of the level of integration ($j_{PCi_n}$) ranging from 0 to 3, where 0 denotes a weak level of integration and 3 denotes a very strong level of integration.

In addition, weights ($k_n$) were assigned to each line of interaction by expert method in terms of their impact on the efficiency of development investment activities. Since further the study analyzes the productivity of the construction industry as an indicator of efficiency, the increased level of influence will be typical for the lines of interaction of entities characterized by the greatest amount of mutual effects. As we know, these areas include relationships with designers, contractors, as well as suppliers of materials and construction machinery and equipment. These areas are assigned boosting weight of 2. In the line of international cooperation, the level of influence is estimated at 0.5, due to the fact that the construction market is characterized by significant territoriality. Thus, the final coefficient of the level of integration of the development investment process in the $i^{\text{th}}$ country is calculated by summing up specific indices taking into account the boosting weights [formula (1)].

$$J_{PCi} = \sum\nolimits_{n=1}^{6} j_{PCi_n} * k_n \qquad (1)$$

A correlation analysis of the dependence of the index of the level of integration of the development investment process subjects (a feature factor) and the efficiency of construction activity (quantified through the average performance indicator, assessed as the ratio of the total turnover of the construction industry to the number of employed in the industry) (resulting factor) was also carried out. The $r_{yx}$ pair correlation coefficient is calculated; the linear regression equation was not compiled due to the fact that the available empirical materials did not allow to build models with a sufficient level of significance (due to the small number of observations) that could be used as a predictive tool.

## 2.2 Mechanisms and Models for Coordinating the Activities of Economic Agents

Economic theory identifies three main mechanisms for coordinating the activities of economic agents: market, organizational, and public, the use of each of which is associated with certain institutional costs, manifested to a greater or lesser extent depending on the specific conditions of the project implementation. According to the Coase theorem, the market mechanism of interaction (the conclusion of voluntary agreements and contracts) is most effective in the case of small transaction costs and clear specifications of property rights [7].

The available research materials for market mechanisms for the interaction of economic agents indicate the concentration of attention of the scientific community in terms of approaches to coordination of goals, interests, operation of independent subjects of the joint project [6, 9–11]. As essentially developed tools we can single out mechanisms for managing relationships based on the allocation of tasks, complementarity of resources, orchestration of joint activities, etc. Common to these approaches is the focus on subordination of such attributes of interactions between entities as distributed tasks, areas of responsibility, results obtained, etc. to the single goal of the project. These tools practically do not contain economic mechanisms for fixing the desired set of attributes, as they are based mainly on the ideas of relationship, involvement, moral obligations and responsibilities in relation to society of the project subjects. Researchers point out that the distribution of costs and benefits is carried out on the principles of trust. The paper

[4], which contains models of economic interaction between subjects, the methodological basis of economic partnership within the framework of a development investment project, a matrix of costs and benefits, as well as the conditions for their obtaining by differentiated types of project subjects, seems somewhat oppositional in these conditions.

The digital basis for the integration of independent design entities on the basis of the market mechanism of interaction is represented by the technology of information modeling of the development investment project based on the BIM principles (BIM, Building Information Modeling), which in a narrow sense [for example, 1, 6, 12, 13] is a new approach to architectural and construction design, which boils down to creation of a computer model of a building or a structure. Within this consideration, BIM is essentially a new digital technology for the design of buildings, which has arisen in response to the multiplying and ever increasing amount of information that needs to be taken into account at the design stage.

### 2.3  Data on the Effects of Information Modeling Technology Implementation

Current research on the implementation of information modeling technologies [3, 14–17, 21, 22, 25] reveals not only and not so much the effects at the design stage, but also throughout the life cycle of the project due to the following major benefits:

- at the design stage by eliminating losses caused by missing design deadlines (estimated up to 40% [21]);
- at the construction stage by reducing the time of the facility's construction as a result of reduced requests for changes in design documentation, optimization of work schedules and tender procedures, by reducing the cost of construction as a result of more accurate planning of materials and equipment, reduction of waste, failures and alterations (estimated up to 37%, [21]), by improving the quality of the facility created as a result of automated detection and avoidance of collisions (inaccuracies, errors and inconsistencies) already at the design stage;
- at the operation stage by reducing the cost of operation and maintenance of the facility, improving its safety and reducing the impact and burden on the environment;
- at the liquidation stage by reducing the cost and timing of redevelopment.

Moreover, the transition to BIM in the field of design also creates synergistic effects, when, for example, in the process of urban and territorial development, the cost and timing of changes in the social, engineering and utilities, etc., space is reduced, the safety and environmental friendliness of the facility increases, the burden on the environment is reduced, the consumption of resources in other projects and programs, as well as at other facilities, is reduced.

Studies also show that the design BIM models solve only part of the tasks of the subjects of the construction and, even more so, of the operation stage, as it is a necessary instrument at these stages, but by no means a sufficient one. During the construction of a facility, new information is generated, which constantly requires storage, transfer and application by users. This information is already linked into the construction information model, which fully facilitates the objectives of the subjects of the construction stage. After the object is commissioned into operation (which, by the way, also requires its own information model), the processes of maintenance and operational support of the facility, as well as interaction with the environment, are launched, which forms a new flow of information that is already dynamic in nature.

### 2.4 Institutional Conditions for the Implementation of Information Modeling Technology in Russia

An analysis of the management of the digital transformation of the market economy [for example, 17, 23, 24] has shown that the following steps have now been taken to create the conditions for the effective implementation and use of information modelling technology in the development investment process in the Russian Federation:

- the President of the Russian Federation instructed to introduce information modeling in the construction industry (2018).
- the solutions to the digital transformation of the construction industry are formulated in the national "Digital Construction" Project. The draft document contains in full the measures developed by the Russian Ministry of Construction in accordance with the President's instruction (2018).
- the Ministry of Construction and the FSC FAE (Federal Standardization Center), with the participation of the Russian Chief State Expertise, have developed the concept of implementing a system of management of the life cycle of capital construction facilities using information modeling technology (further denoted as Concept), which was significantly improved after discussions with expert groups (2018–2020).
- Amendments to the Russian City Building Code, which legislated concepts such as "information model" and "building information classifier", have established a "seamless" procedure for the exchange of information, documents and materials about the capital construction facility using existing state information systems, as well as the powers of the subjects of urban planning relations to apply information modeling at all stages of the life cycle of the capital construction facility (2019).

## 3  Results

3.1 In the context of the contractual separation of the development investment project, independent entities maximize the economic impact based on market conditions. At the same time, the results of studies of the nature of effects formations within and between the stages of the development investment project demonstrate the presence of a large number of mutual effects. This means that the maximum full economic effect of investment will occur further than the maximum economic effect manifesting within the stage. In the absence of coordination and integration, there will be a loss of efficiency of individual stages and of the development investment project as a whole due to the lack of optimal values in the level of investment.

The nature of the relationships between the actors of different stages of the development investment process is best reflected in the situation of "as is" comparison with imagining each project team (designers, contractors, customers, investors, etc.) in "their boat" with a specific level of "operational qualities" (i.e. technologies and resources) that, although moving towards the "one shore" of the target field of the project, but each towards its own "berths" characterizing individual goals, uncorrelated with the integral efficiency of the project. With this nature of the relationship, most participants have no interest in the effectiveness of other actors in the project, so in the absence of additional incentives they are not ready to "take in tow", that is, to spend their own resources without getting a return at the appropriate stage of the project. The target situation "as it should be" is to place the entire team of entities on one "big ship" seeking a single "berth" of the target of the development investment project, the achievement of which ensures its integral efficiency. At the same time, due to the involvement of synergetic potential, the "operational qualities of the ship" (resources, technology, etc.) are much higher than those of individual "boats".

The above theoretical justification for the integration of development investment entities is also confirmed by the results of empirical verification. Table 1 contains the results of the quantitative analysis of the level of integration of the development investment process in the country, including in correlation with the average performance of construction activity. The results suggest that there is a clear statistical trend in the very strong relationship between the Chaddock scale under study. In other words, the results demonstrate that countries with the strongest performance in the construction industry have higher rate of efficiency in the inter-subject integration.

**Table. 1.** Results of quantitative analysis of the level of integration of development investment actors in the country and the average productivity of construction activities

| Group of subjects | Contractors and suppliers | Designers | Consumers | State | Universities | International cooperation | Specific indices of interaction degree of the $n^{th}$ group taking into account the weight if the $i^{th}$ country, $J_{PCn} * k_n$ | | | | | | Resulting interaction degree index in the $i^{th}$ country, $J_{PCi}$ | Average performance of the construction industry in the $i^{th}$ country, $W_i$ | Correlation coefficient, $r_{yx}$ |
|---|---|---|---|---|---|---|---|---|---|---|---|---|---|---|---|
| Weight coefficient, $k_n$, $n = [1;6]$ | 2 | 2 | 1 | 1 | 1 | 0.5 | | | | | | | | | |
| Country | Specific indices of the interaction degree of the nth group in the $i^{th}$ country, $J_{PCn}$ | | | | | | | | | | | | | | |
| Denmark | 3 | 2 | 2 | 3 | 0 | 0 | 6 | 4 | 2 | 3 | 0 | 0 | 15 | 8.09% | |
| Sweden | 2 | 0 | 3 | 0 | 3 | 1 | 4 | 0 | 3 | 0 | 3 | 0.5 | 10.5 | 6.55% | |
| Germany | 2 | 0 | 0 | 0 | 1 | 1 | 4 | 0 | 0 | 0 | 1 | 0.5 | 5.5 | 5.87% | 0.9404 |
| UK | 0 | 0 | 2 | 2 | 1 | 1 | 0 | 0 | 2 | 2 | 1 | 0.5 | 5.5 | 4.41% | |
| France | 1 | 0 | 1 | 0.5 | 0.5 | 1 | 2 | 0 | 0.5 | 0.5 | 0.5 | 0.5 | 4 | 4.39% | |

3.2. In the development investment management, the main model of implementation of development investment projects on the basis of the market mechanism of coordination of entities we can single out IPD-partnership contracting (IPD, Integrated Project Delivery). Within such partnerships, specific processes, as well as interests and competencies of its subjects, are aggregated into a single process subordinated to the achievement of the overall goal of the project and ensuring its final efficiency, through the formation of a single team of key participants in the development investment project (IPD subjects). The market mechanism for the integration of entities on the basis of the IPD partnership is implemented based on the *open book* principle, implying transparency of the formation of costs, results and risk management of the project. The interaction of

IPD subjects is carried out through the conclusion of a multilateral IPD contract, which solidifies their functions, rights, obligations, as well as mechanisms and procedures for the distribution of costs and benefits. The distribution of costs and benefits within an IPD project based on the principles of trust is a very weak basis in conceptual terms. Ensuring the interaction of the subjects, when the effects of an individual entity are dependent, among other things, on the investments of others, and integral efficiency, among other things, on additional synergistic effects, is facilitated by the entire development investment cycle, requires a more stringent mechanism. The interaction of subjects under the IPD contract maximizes the effect of a development investment project through leveling the action of externalities through internalization and as a result of conflict avoidance, shortening the duration of the stages, reducing transaction costs for disputes, courts, etc. Such projects provide for management, which is implemented mainly through anchor enterprises through the so-called soft dominance, which ensures that the interests and actions of the project participants are synchronized within the project to achieve its common goal.

3.3. The external nature of most of the effects of the implementation of information modeling technology in the development investment process determines the need to use internalizing approaches, which involve the presence of the long-term end user of *BIM* products of all stages of the project, in the absence of which the introduction of information modeling technology does not reach effective volumes (Fig. 1). As the ultimate long-term user of *BIM* products of all stages of the project, ensuring the internalization of mutual commercial effects arising during the development investment cycle, we can consider the anchor party of the project from among customers, investors, etc. As the end long-term user of *BIM* products of all stages of the project, ensuring the internalization of externalities in all areas of sustainable development, we can consider the state as a representative of the public sector.

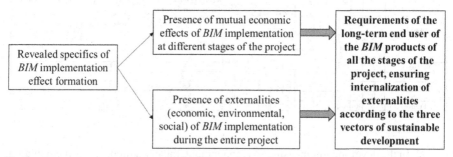

**Fig. 1.** Economic requirements of internalization of externalities of implementation of information modeling technology in the development investment process

The results of the analysis of existing studies in the implementation and development of information modeling technology as a digital integrating basis of the development investment process also demonstrate that, first, maximizing the effects of the *BIM* approach in addition to the implementation of design solutions necessitates the formation of construction and operation *BIM* models, and second, *BIM* allows to form a single information space of the project, which facilitates effective interaction of its independent subjects. Information generated and stored in the *BIM* environment can be jointly used by multiple users (parties, specialists) throughout the project, from its design to the final stage; all changes are recorded and available at any given time. These areas determine the need for a unified software for the formation of *BIM* models of all stages and levels of a development investment project, as well as for standard technologies for internal and external data exchange between various digital models, the development of which is still in its initial stage. These developments, requiring large-scale work and investments, coupled with significant externalities, are associated with the need for the participation of the internalizing public side (Fig. 2).

The results show, on the one hand, that information modeling technology can become a key integrating tool of the development investment process through the *IPD* contracting system, and, on the other hand, extensive work on the development of software and data-sharing BIM products of all design stages and levels is required.

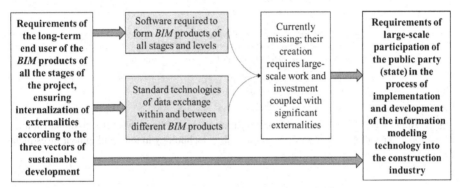

**Fig. 2.** Technological requirements of *BIM* model implementation within the development investment process and justification of the involvement of the public party

3.4. The principles of the development and use of *BIM* dictate the need for large-scale public involvement in this process. Scientific findings indicate that a purely commercial route is associated with a significant number of externalities, which makes it, in fact, unrealizable. The implementation of the *BIM* approach achieves its full effect only if it is systemically applied throughout the entire industry (Fig. 3).

In addition to creating an institutional environment for the new system of interaction between the parties of the development investment process based on the *BIM* principles, which requires complete restructuring of the current legislation and technical standards, it is necessary to create an up-to-date system of management of *BIM* products throughout their entire life cycle in a single information space, that is, a complex BIM platform including all the necessary content (libraries, databases, classifiers, navigators for contractors, suppliers and operators with the necessary ratings, etc.), as well as to facilitate operation of *BIM* providers, that is, entities on a professional basis providing access to a *BIM* platform on the principles of software- and IT-sharing, as well as carrying out storage, transfer and inheritance of the complex of *BIM* products. The result of the introduction of the *BIM* approach for the whole range of development investment actors should be a systemic change in the organization of their interactions within the project

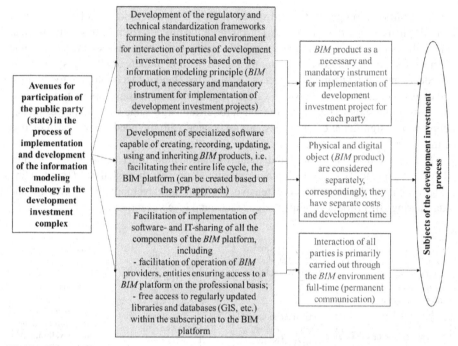

**Fig. 3.** Approaches and results of the participation of the public party in the process of implementation and development of information modeling technology in the development investment complex

in the direction of permanent communications in the *BIM* environment based on a high level of collaboration. Only under these conditions will there be a full digital base of a successful IPD partnership.

In the context of these communications, the physical and digital (BIM product) objects created and used in the course of the development investment project are considered separately. The quality of management of the creation and operation of the digital object, similarly to the physical object, is characterized by the cost and timing of its creation, as well as the quality of the *BIM* product, which form its value, which, in turn, ensures the quality of management of the formation and use of a physical object, that is, the reduction of its cost, time required, as well as improvement of the quality in the broadest sense of the term. Ideally, a physical object formed as part of a development investment project, as well as the processes of its subsequent operation, should be the result of the best and most effective solutions for digital BIM products in a BIM environment.

## 4   Discussion

4.1. Analysis of the results obtained confirms the hypothesis of the effectiveness of the integration of the major parties in the development investment process, which correlates with the provisions of a wide range of scientific publications [in particular, 1–3, 8, 18, 22].

The obtained on the basis of empirical research [8] quantitative assessments of the relationship between the level of integration and performance (efficiency) of construction activities are explained, in the first place, by eliminating a wide range of mutual effects [22], and the fact that, in a more integrated environment, a variety of innovations are being invested in more intensively, for example those aimed at knowledge transfer [18] and supply chain technologies [3], which facilitates a significant improvement in building products or processes and ultimately the efficiency of the entire development investment project. These findings are also supported by the research on the dissemination of innovation in the construction industry, taking into account the context of digitalization [19, 20].

4.2. As an instrument of the market mechanism for the interaction of development investment parties, as the most effective in a situation of insignificant transaction costs, we are investigating the IPD approach, the advantage of which was generally recognized [13, 22, 26, 27]. Existing methods of coordinating the subjects of joint projects (which include IPD contracts) are based mainly on the principles of trust, mutuality, involvement and moral obligations [6, 9–11]. We conclude that these principles are a very weak conceptual basis in terms of the distribution of costs and benefits of the IPD project. When the effects of an individual subject depend, among other things, on the investments of others, and integral efficiency is provided by the entire development investment cycle through leveling the influence of externalities, a more stringent economic mechanism is required.

4.3. In contrast to the simple aggregation of IPD and BIM approaches (such as in the paper [13], which analyzes the presence of synergies in their joint implementation, and in the [22, 27]), we conclude that the effective level of development of information modeling technology based on the BIM principles ensures the implementation of a full-fledged economic IPD partnership. The analysis of results of current research in the area of economic development considering the digitalization of information and communication technologies, as well as the existing practice of applying BIM concepts in the construction industry demonstrates its ability to create a single digital information space of the project [for example, 25, 26]. This makes it possible to reduce transaction costs to values close to zero. As we know, the insignificant value of transaction costs forms efficient conditions for conclusion of voluntary agreements and contracts, including in the format of the IPD partnership. From the above considerations it logically follows that information modeling technology will become the integrating basis of the development investment process, ensuring its effectiveness.

4.4. The results indicate that the effectiveness of information modelling technology implementation in the development investment process will be achieved in the case of a much broader approach than its use only as a digital tool of architectural and construction design. This understanding of information modeling technology is present in a fairly wide range of modern scientific publications [1, 12, 13, 28]. The broad approach we are implementing, supported by a number of studies [3, 15], involves micro-level consideration of BIM as a new technology for managing a development investment project based on the principles of digital information modeling throughout its life cycle, and at the mesoeconomic level, as a new system of relationships (organizational, economic, etc.) between all construction industry actors based on approaches and principles of digital information modeling and management. At the same time, even within the broad consideration, BIM is regarded as one of the technologies of the Industry 4.0 within the construction industry (along, for example, with cloud computing) [2, 3, 29]. In other words, BIM is perceived mainly as a technological innovation, i.e. attempts are made to digitize existing business processes and relationships of entities. We propose to consider the information modeling technology, first of all, as an organizational innovation, involving qualitative changes in the existing organizational and management models, which, due to significant externalities, is possible only with substantial government support.

## 5 Conclusion

Effective implementation of information modeling technology requires systemic transformations of the development investment economy, as well as existing organizational and management models in the direction of creating a single information space of the industry and permanent interaction of its subjects, ensuring the increase in the value of development investment projects taking into account the growth of the effects of individual participants. In turn, for the systemic transformation of the industry, it is necessary to create a single information space based on software and data exchange technologies, the creation of which should be largely funded with governmental participation. In these conditions, the need for entities providing access to the BIM environment to function on a professional basis, as well as storing, transferring and inheriting a complex of BIM products, is no longer doubted. As there is no doubt about the need for the existence of classic Internet providers or cellular operators, who also provide access to information resources for a certain fee.

It begs the conclusion that the professional community of the construction industry will become a supporter and partner of the introduction of full-fledged information modeling technology only in the case of the systemic application of the BIM approach throughout the industry, with the large-scale participation of the state, ensuring the internalization of the externalities of the introduction of BIM for individual private participants. Only under these conditions will a full-fledged digital base be laid, integrating the subjects of the development investment process within the framework of the market principles of IPD partnership.

## References

1. Pruskova, K.: Beginning of real wide us of BIM technology in Czech Republic. IOP Conf. Ser. Mater. Sci. Eng. **471**, 102010 (2019)
2. Maskuriy R., et al.: Industry 4.0 for the construction industry—how ready is the industry? Appl. Sci. **9**(14), 2819 (2019)
3. Dallasega, P., Rauch, E., Linder, C.: Industry 4.0 as an enabler of proximity for construc-tion supply chains: a systematic literature review. Comput. Ind. **99**, 205–225 (2018). https://doi.org/10.1016/j.compind.2018.03.039
4. Kornilova, S.V.: Market mechanism for ensuring the effectiveness of development investment projects. Econ. Manage. **26**(11)(181), 1263–1270. https://doi.org/10.35854/1998-1627-2020-11-1263-1270

5. Kvasha, N.V., Demidenko, D.S., Voroshin, E.A.: Industrial development in the conditions of digitalization of infocommunication technologies. St. Petersburg State Polytech. Univ. J. Econ. **11**(2), 17–27 (2018). https://doi.org/10.18721/JE.11202

6. Ndubisi, N.O., Dayan, M., Yeniaras, V., Al-hawari, M.: The effects of complementarity of knowledge and capabilities on joint innovation capabilities and service innovation: the role of competitive intensity and demand uncertainty. Ind. Market. Manage. **89**, 196–208 (2020). https://doi.org/10.1016/j.indmarman.2019.05.011

7. Coase R.H.: The problem of social cost. In: Gopalakrishnan, C. (eds.) Classic Papers in Natural Resource Economics. Palgrave Macmillan, London (1960). https://doi.org/10.1057/9780230523210_6

8. Miozzo, M., Dewick, P.: Networks and innovation in European construction: benefits from inter-organisational co-operation in a fragmented industry. Int. J. Technol. Manage. **27**(1), 68–92 (2004). https://doi.org/10.1504/IJTM.2004.003882

9. Furlotti, M., Soda, G.: Fit for the task: complementarity, asymmetry, and partner selection in alliances. Org. Sci. **29**(5), 755–987 (2018). https://doi.org/10.1287/orsc.2018.1205

10. Milwood, P.A., Roehl, W.S.: Orchestration of innovation networks in collaborative settings. Int. J. Contemp. Hosp. Manage. **30**(6), 2562–2582 (2018). https://doi.org/10.1108/IJCHM-07-2016-0401

11. Linde, L., Sjödin, D., Parida, V., Wincent, J.: Dynamic capabilities for ecosystem orchestration a capability-based framework for smart city innovation initiatives. Technol. Forecast. Soc. Change **166**, 120614 (2021). https://doi.org/10.1016/j.techfore.2021.12061

12. Ceccon, L., Villa, D.: AI-BIM interdisciplinary spill-overs: prospected interplay of AI and BIM development paradigms. In: Bolognesi, C., Villa, D. (eds.) From Building Information Modelling to Mixed Reality, pp. 195–217. Springer, Cham (2021). https://doi.org/10.1007/978-3-030-49278-6_12

13. Nguyen, P., Akhavian, R.: Synergistic effect of integrated project delivery, lean construction, and building information modeling on project performance measures: a quantitative and qualitative analysis. Adv. Civil Eng. (2019). https://doi.org/10.1155/2019/1267048

14. Oesterreich, T.D., Teuteberg, F.: Looking at the big picture of IS investment appraisal through the lens of systems theory: a system dynamics approach for understanding the economic impact of BIM. Comput. Ind. **99**, 262–281 (2018)

15. Woodhead, R., Stephenson, P., Morrey, D.: Digital construction: from point solutions to IoT ecosystem. Autom. Constr. **93**, 35–46 (2018)

16. Churbanov, A.E., Shamara, Yu.A.: Influence of information modeling technology on the development of the development investment process. MSUCE Herald. **13**(7)(118), 824–835 (2018). https://doi.org/10.22227/1997-0935.2018.7.824-835

17. Shevelev, A., Novikov, A., Budagov, A., Zhulega, I.: Digital transformation of market economy: institutional approach. In: В сборнике: European Proceedings of Social and Behavioural Sciences EpSBS, pp. 1278–1284. Krasnoyarsk Science and Technology City Hall. Krasnoyarsk (2020). https://doi.org/10.15405/epsbs.2020.10.03.147

18. Owusu-Manu, D.-G., John Edwards, D., Pärn, E.A., Antwi-Afari, M.F., Aigbavboa, C.: The knowledge enablers of knowledge transfer: a study in the construction industries in Ghana. J. Eng. Des. Technol. **16**(2), 194–210 (2018). https://doi.org/10.1108/JEDT-02-2017-0015

19. Kuladzhi, T., Babkin, A., Murtazaev, S.A.: Matrix tool for efficiency assessment of production of building materials and constructions in the digital economy. Adv. Intell. Syst. Comput. **692**, 1333–1346 (2017). https://doi.org/10.1007/978-3-319-70987-1_141

20. Rodionov, D., Nikolova, L., Abramchikova, N., Velikova, M., Mazuba, K.R.J.: Development of the analysis model of innovative projects efficiency management in the context of digitalization. In: Proceedings of the 2nd International Scientific Conference on Innovations in Digital Economy: SPBPU IDE-2020, pp. 1–7, October 2020

21. Trofimova, L.A., Trofimov, V.V.: Implementation of the strategy of innovative development of the construction industry in the Russian Federation on the basis of information modeling of industrial and civil facilities 1(05), 31–35 (2017). https://doi.org/10.18454/mca.2017.05.1

22. Piroozfar, P., Farr, E.R.P., Zadeh, A.H.M., Inacio, S.T., Kil-gallon, S., Jin, R.: Facilitating building information modelling (BIM) using integrated project delivery (IPD): a UK perspective. J. Build. Eng. **26**, 100907 (2019). https://doi.org/10.1016/j.jobe.2019.100907

23. Korshunova, E.M., Taranukha, N.L., Borshova, P.A., Korshunov, A.F.: Factors limiting the use of modern BIM technologies in reconstruction projects. Intell. Syst. Prod. **18**(3), 53–57 (2020). https://doi.org/10.22213/2410-9304-2020-3-53-57

24. Budagov, A.S., Sukhova, N.A.: Problems of effective business digital transformation management. In: European Proceedings of Social and Behavioural Sciences EpSBS, Krasnoyarsk, 20–22 мая 2020 года/Krasnoyarsk Science and Technology City Hall. Krasnoyarsk: European Proceedings, pp. 428–437 (2020). https://doi.org/10.15405/epsbs.2020.10.03.48.

25. Marinho, A., Couto, J., Teixeira, J.: Relational contracting and its combination with the BIM methodology in mitigating asymmetric information problems in construction projects. J. Civil Eng. Manage. **27**(4), 217–229 (2021). https://doi.org/10.3846/jcem.2021.14742

26. Ahmad, I., Azhar, N., Chowdhury, A.: Enhancement of IPD characteristics as impelled by information and communication technology. J. Manage. Eng. **35**(1) (2019). https://doi.org/10.1061/(ASCE)ME.1943-5479.0000670

27. Salim, M.S., Mahjoob, A.M.R.: Integrated project delivery (IPD) method with BIM to improve the project performance: a case study in the Republic of Iraq. Asian J. Civil Eng. **21**(6), 947–957 (2020). https://doi.org/10.1007/s42107-020-00251-1

28. Darko, A., Chan, A.P., Yang, Y., Tetteh, M.O.: Building information model-ing (BIM)-based modular integrated construction risk management–critical survey and future needs. Comput. Ind. **123**, 103327 (2020). https://doi.org/10.1016/j.compind.2020.103327

29. Demirkesen, S., Tezel, A.: Investigating major challenges for Industry 4.0 adoption among construction companies. Eng. Constr. Archit. Manage. (2021, ahead-of-print No. ahead-of-print). https://doi.org/10.1108/ECAM-12-2020-1059.

# Automation of the Business Process to Interacting with Clients of a Company Based on Bitrix24 (on the Example of a Software Development and Implementation Company)

Mariya Volik[1]([⊠]) [iD] and Tatyana Kopysheva[2] [iD]

[1] Financial University under the Government of the Russian Federation,
Molodezhnaya Street, 7, 362002 Vladikavkaz, Russia
[2] Chuvash State University named after I.N. Ulyanov, Moskovsky prospect, 15,
428015 Cheboksary, Russia

**Abstract.** In modern economic conditions, the issue of development and stable activity of each company is relevant. Many companies interact with clients on the sale of goods, products, services. To increase the customer base and its retention, it is necessary to use modern digital technologies. The most popular and affordable for small companies are network and mobile technologies. In this work, a preliminary project has been developed for automating the business process of interacting with clients of the company for the development and implementation of software. A preliminary analysis of the company's activities and the automated business process showed that the implementation of a Bitrix24-based CRM system is required. In this regard, a description of the project for the implementation of a CRM system has been prepared, reflecting the advantages of automating interaction with customers. Also, a preliminary calendar plan has been developed, consisting of four stages and a complex of subtasks. A preliminary assessment of the amount of investment for the implementation of the project has been carried out. The implementation of the proposed project will optimize the work of client managers, reduce the number of errors, make the interaction between clients more transparent, and optimize the interaction between employees and departments. Automation of interaction with customers of the company will allow in the future to reduce current costs and increase net profit. The proposed project is universal for different companies and, after making the necessary adjustments, can be used with other initial data.

**Keywords:** Information system · Information technology · Business process · Customer Relationship Management · CRM · Bitrix24

## 1 Introduction

In modern economic conditions, the development of all areas of activity is of great importance. The main parameters of activity management are associated with a multi-structure (state, cooperative and private sector). Automation of production is associated

© Springer Nature Switzerland AG 2021
D. Rodionov et al. (Eds.): SPBPU IDE 2020, CCIS 1445, pp. 73–85, 2021.
https://doi.org/10.1007/978-3-030-84845-3_5

with the use of automatic and/or automated systems that make it possible to completely or partially release labor resources from the operations performed. However, it is necessary to automate not only production business processes, but also business processes related to company management, interaction with suppliers and customers.

Currently, improving the quality of enterprise management (management decisions, developing plans, methods of manufacturing products and providing services) is possible only with the direct use of modern information technologies (IT), which help to achieve the set goals by automating the necessary business processes. Implementation of IT provides different options for implementation and strategic opportunities. [23, 24] Automation of business processes implies a set of actions to analyze the current state, develop an optimization plan, select and implement modern information technologies, and train end users. As a result of automation, the company's management solves the tasks of reducing the amount of routine work and the number of errors, increasing labor productivity, adding new opportunities for information processing, increasing competitiveness and increasing the market value of the business. That is, the automation of business processes is necessary not only in cases of reorientation to new tasks, but also to improve the principles of enterprise management and interaction with customers. Scientific and practical researches of different authors are devoted to the solution of such a problem Abbad Andaloussi A., Burattin A., Kindler E. and etc. [1], Guerola-Navarro V., Gil-Gomez H., Oltra-Badenes R., and etc. [9], Ibragimova R. and etc. [10].

A review analysis of literary sources has shown the relevance of research in the field of automation of business processes of a company in order to increase the company's profits, enhancements the level of its competitiveness, reduce costs, etc. In this work, the concept of automation on interaction with clients based on Bitrix24 has been developed on the example of a company engaged in the development and implementation of software (software).

## 2 Literature Review

Rapidly developing information technologies have an increasing impact on many areas of human activity, including economic. In this regard, the methods of effective management of companies using information systems and information technologies are widely used, which make it possible to keep under control the increasing external and internal flows of information necessary for analysis, making forecasts and making management decisions. However, in the area under study, the situation is not well defined.

First of all, this is due to the constantly increasing number of innovative proposals that require serious investment. In addition, the costs of developing the information technology infrastructure of companies are increasing significantly. There is a change in the importance of modern IT in the business of companies. Modern IT has ceased to perform an auxiliary function. They are the most important component of the activities of each company in terms of optimizing the business processes of the enterprise based on end-to-end automation of their constituent business functions. However, before any way to improve the business processes of any company, it is necessary to analyze its activities, highlight and describe the existing business processes and human resources. And only after a set of measures has been taken, it is necessary to start developing measures, models and/or projects to improve business processes, depending on the scale of the company and the goals pursued within the framework of the business strategy. For the solution of most problems of this kind, appropriate methods, methodologies, technologies, including software, have been developed and are widely used [18, 21].

The main goal of business process management is to improve processes in accordance with the strategic goals of the company. Each business process must be organized in such a way that the achievement of the company's business goals is ensured.

Currently, to manage business processes (BPM, Business Process Management) modern application programs are used, which make it possible to simplify the management of various business processes of the company. In practice, not every company needs a complete set of software modules. Each company selects the modules that are required to automate business process management: graphical modeling; modeling the dynamics of a business process; development of applications for creating an interface; monitoring of business processes; control modules, etc. Thus, for the effective functioning of an enterprise, it is necessary to identify numerous types of activities (business processes); identify all their relationships and interactions; manage interrelated processes as a consistent system. The description of business processes will allow to analyze the need for their improvement or reengineering [14, 26].

Most companies in the course of their activities interact with customers. In this regard, it is advisable to automate this particular business process. An effective customer engagement strategy is to develop customer-driven solutions rather than finding customers for outdated solutions. This approach is aimed at building loyalty, strengthening the customer base, and increasing the company's success.

Thus, Customer Relationship Management (CRM) is a strategy that focuses on creating profitable and lasting relationships with consumers based on their individual needs. When organizing CRM, areas directly related to customers are affected: sales, service and marketing management. The use of CRM provides the company with the following benefits: increased efficiency of employees works; decrease in the number of lost clients; reducing the cost of marketing activities; elimination of leakage of customer information; reduction in management costs [11, 25].

In the context of automation of interaction with customers, application software is used to collect and store information about customers, control and analyze the results of interaction with them. Modern IT is used at different stages of working with clients and allows you to increase the speed of the company's business processes, which has a positive effect on profits.

Currently, CRM systems are popular in the service and trade sectors. However, the tasks that companies solve with the help of application systems for interaction with customers are relevant for other areas of activity. CRM systems are actively used in the financial, insurance, telecommunications, industrial, construction industries, IT and consulting [4, 6].

The main goal of the implementation and use of CRM is to optimize and automate the business processes of customer relationships based on the effective management of the relevant information flows. The result is attracting and retaining profitable customers, as well as building mutually beneficial and long-term relationships with them.

The Russian market for CRM systems is actively developing, and import substitution makes various applied solutions more relevant. The rating of the best CRM systems for business is represented by the following developments: Megaplan; Bitrix24; CRM «Simple Business»; Sales Creatio; Microsoft Dynamics CRM; amoCRM; Mango CRM; Wrike; Trello; SugarCRM, etc. [3, 12].

The Bitrix24 service is a corporate portal that combining the functions of social networks, project management, tasks, personnel and interaction with clients. There are two options (tariffs) of work: a cloud solution (purchase of a SAAS solution, in which access to remote servers of Bitrix24 is paid for) and a boxed solution (purchase of a Stand-Alone software product, in which software is purchased for installation on the company's own server) [6, 15].

CRM Bitrix24 is one of the most popular application solutions at the present time. The use of the system will allow the company to solve a number of problems in the automation of work within the company and interaction with clients. The implementation of Bitrix24 is aimed at bringing the company's business processes to a new topical level [17, 20].

CRM Bitrix24 provides the management and employees of the company with the following main features:

- clear presentation of requests in processing;
- reflection of the entire history of the deal in the CRM card;
- creating quick chats to discuss deals, leads, companies, contacts;
- registration and recording of call conversations in CRM;
- sending letters to clients from the application;
- distribution of all messages from CRM among managers;
- sales automation;
- distribution of the sales plan among employees;
- creating repeat deals with clients and transferring them to the manager for work;
- organization of sales in chats of social networks and instant messengers;
- paperwork in CRM;
- bi-directional data exchange (goods, contacts, companies, orders) with 1C in real time;
- CRM marketing;
- quick and easy setup of a sales funnel between different stages;
- ready-made reports on sales and the work of managers [8, 16, 19].

The advantages of Bitrix24 include:

- A wide range of possibilities. The system greatly simplifies the scheduling of tasks and maintenance of the client base, makes it possible to ensure integration with the online store and automate the business process;
- Possibility of integration with 1C. This is a very important parameter for every company. Correct setting of Bitrix24 helps to solve the mass of tasks;
- Simplicity. It is easy to work with the system, and the process of training an ordinary employee takes little time, and it is also possible to quickly search for the necessary information [5, 13].

Disadvantages of Bitrix24:

- The need to work within the functionality established by the developers. In terms of customization, the system is closed, so it is not so easy to expand customer data fields or carry out similar work;
- The need to get used to the interface. Many users note that it is far from immediately possible to navigate in all the wealth of possibilities. Bitrix24 has a rather complex visualization and it is not always easy to find the desired task among the completed ones [2, 7].

Thus, the Russian CRM Bitrix24 is popular with companies in various fields of activity. Automation of business processes of interaction with customers using Bitrix24 tools has a number of problems. However, professional implementation and configuration of the system will allow the company to use all its advantages and capabilities. And the maximum transparency of the company's business processes will lead to an raising in the efficiency of activities, and increase hence the company's profit. In this regard, before the implementation of CRM Bitrix24, it is advisable to research the company's activities, prepare a description and models of automated business processes, and develop an up-to-date implementation project.

## 3   Research Methods

Business process management implies the need to model them to facilitate analysis and further development (improvement). Based on the results of such a study, the company's management needs to ensure the development of a project for automation, improvement or reengineering of business processes.

An obligatory stage in the management of an automation project is its planning. The project plan is developed based on the project charter and is organized information that can be used to plan, organize and control the activities of the working group. As a result of the formation of the project schedule, a clear understanding of the timing of achieving the project goals and key events in the project is provided; determination of the required amount of resources (labor, financial, etc.); formation of a list of tasks aimed at achieving the goals of the project, and their timing; streamlining the processes of information exchange between project participants; linking different areas of project activities, presentation of project work in a single format [16].

For effective project management, Microsoft Project tools can be used, which provides convenient collaboration capabilities that allow you to quickly start and successfully complete projects. MsProject is designed to assist the project manager in developing plans, allocating resources to stages, tracking progress, and analyzing scope of work. The developed project for the implementation of a system of interaction with clients using MsProject will allow the company's specialists, before its implementation, to make the necessary changes and additions to the proposed stages and subtasks, start and end dates of work, deadlines for work, used resources (labor resources, material, costs), etc. etc. MsProject provides the ability to monitor the implementation of the stages and work of the project [26, 27].

In this paper, based on the analysis activities of the IT company carried out in [27], has been prepared a preliminary project for the automation of interaction with the company's customers (the company name is not given due to the preservation of the confidentiality of information) on the basis of a CRM system. This study was initiated by the company's management in order to improve the efficiency of the activities company.

The main activities of the investigated company: design and development of software; turnkey website development and creation of an online store; site placement on the Internet; development of modules for the popular CMS DotNetNuke; Backend development for mobile applications and web services; registration and renewal of domain names; registration of computer programs and databases; conducting training events; development of regulatory and technical documentation. The main goals of the company in the field of quality are: a significant increase in business manageability and ensuring transparency of activities to increase customer satisfaction, increase capitalization and attract investors, the prosperity of the company and the growth of the well-being of each employee. In this regard, the company's management pays special attention to improving the efficiency of interaction with customers.

## 4   Research Results

The work [27] an analysis of the activities of the company under study was carried out, the results of which showed that in order to improve the system of interaction with customers of the company, it is advisable to develop a project for the introducing of CRM based on Bitrix24. In this study, a preliminary draft of measures to be taken to reduce risks and unreasonable costs was developed. As a result of the implementation of the project, it is expected to increase the productivity of employees; ensuring effective and transparent interaction with clients; quality control of work with clients, analysis of satisfaction; control of the return of financial documents, prevention of problems. The computers of the investigated company are connected to a local network and have Internet access. All the necessary equipment is available to work with clients, which does not require replacement or modernization. Therefore, the project provides only for the purchase, installation and configuration of software, as well as training for employees.

The start date of the project was set for the beginning of the 3rd quarter of 2020. However, this date was postponed by the management of the company indefinitely due to the financial condition. In addition, the timing of the work is provided in such a way that it does not negatively affect the rest of the company's business processes. The proposed project takes into account the following working time schedule of the implementation group: 5 day working week, working hours from 9 to 18 h with a lunch break from 13 to 14 h. On pre-holiday days, working hours end one hour earlier. Due to the fact that the project will be implemented by the company's employees, additional labor costs for wages will not be required. It is advisable to outsource the installation and configuration of Bitrix24, as well as employee training.

The resources of the project are: time for software implementation, personnel, finances, software, as well as necessary services and costs. The labor resources of the project include: project manager (1 person); working group of the project (3 people); end users participating in training and software testing (4 people). Material resources include software, consumables, and etc. Within the framework of the project, it is planned to purchase only Bitrix24 software. However, in the future, at the discretion of the company's management, the purchase of other machinery, equipment and consumables may be planned.

The proposed project includes four stages:

1. Preliminary work - Creation of a project implementation group; Clarification of information on hardware and software; Formation of ideas about the project, the approaches used, goals, tasks; Clarification of requirements for implementation results; Estimation of terms, resources, types and volumes of work; Assessment of the financial capabilities of the company. Stage duration - 9 working days.
2. Project preparation includes the distribution of powers and responsibilities; definition of organizational and technical requirements; briefing of the working group; analysis of the main directions of costs. Stage duration - 8 working days.
3. The implementation of the project includes clarification of the list and the purchase of the necessary software; installation and configuration of software; user training. Stage duration - 27 working days.
4. Completion of the project consists in software testing; transfer to operation; preparation and signing of an act of completion of work. Stage duration - 5 working days.

Depending on the stages of the project of implementing the system of interaction with clients, subtasks of varying complexity are distinguished, which require solutions during the implementation of the project. In this regard, the stages and subtasks have different labor intensity. At the discretion of the management of the company and the project executors, it is allowed to add, combine, exclude some stages (subtasks) of the project, change their duration based on the results of previously completed work. It is also allowed to start performing some subsequent work simultaneously with unfinished previous ones.

The business goal of the project for the implementation of a system of interaction with clients in the surveyed company is to create a single information space for the company, which will ensure transparency of interaction with clients and between departments, as well as to support the timely formation of documents of a different nature [17, 28].

The project for the implementation of a customer interaction system based on CRM Bitrix24 is intended to improve customer interaction. This will lead to an improved reputation of the company in the eyes of customers. In this regard, at the start of the project, it is necessary to set metrics (indicators, characteristics) by which it will be necessary to determine that the benefits have been achieved. Table 1 shows a map for identifying the business benefits of the project for the implementation of a customer interaction system in the surveyed company based on CRM Bitrix24.

Thus, the implementation of a system of interaction with clients based on CRM Bitrix24 will allow obtaining a number of significant business benefits that can have a positive impact on the company's financial results. The implementation of Bitrix24 will allow you to get a significant effect from the transition to a CRM system with minimal investment in IT infrastructure: ensuring transparency of business processes; improving

**Table 1.** Map for identifying business benefits of the project (compiled by the authors based on research materials).

| Business benefits | The nature of the impact on business | | |
|---|---|---|---|
| | Creating new opportunities | Improving the efficiency of operations | Cancellation of operations |
| Financial | Increasing the customer base | Keeping a customer base | Duplicate operations |
| Quantitative | Increase in the number of clients | Improving the quality of customer interaction | Reducing the number of unaccounted calls |
| Measurable | Expansion of the list of products and services | Increased employee productivity | Reducing the routine of operations |
| Qualitative | Creation of new directions of company development | Improving the quality of customer interaction | Reducing or preventing certain risks |

performance discipline; reducing the time spent by managers and specialists; elimination of information leakage; access to information in a convenient mode; increased competitiveness.

For effective project management of the implementation of a system of interaction with clients, it is necessary to form an organizational structure of the project, which will ensure effective management, planning, execution as planned and at a certain quality level of the stages and corresponding subtasks of the project. The proposed project staffing list includes the following project positions:

– Project Manager - Head of Scientific Research Department (SRD);
– Working group of the project - Engineer (SRD); 1C Specialist (Marketing and Project Support Department - MPSD); Testing Engineer (Product Quality Control Department - PQCD);
– End users - Head of MPSD; Client Relationship Manager (MPSD); General Director (Administrative Office - AO); Executive Director (AO).

To create the organizational structure of the project, it is not required to change the staffing of the company, since during the execution of the subtasks of the project, all involved specialists perform their main duties.

An obligatory stage in the management of an automation project is its planning. As a result of the formation of the project schedule, a clear understanding of the timing of achieving the project goals and key events in the project is provided; determination of the required amount of resources (labor, financial, etc.); formation of a list of tasks aimed at achieving the goals of the project, and their timing; streamlining the processes of information exchange between project participants, etc.

In this work, the project schedule is drawn up using Microsoft Project 2013 tools, which will allow the company's specialists to make the necessary changes and additions before starting the project, as well as monitor the implementation of the project stages and works. All changes that will be made to the project file will allow you to see the results of these changes in real time (for example, the amount of financial costs for labor costs, the amount of financial costs for the acquisition of material resources, of the project end date, etc.). Table 2 presents a project task list.

It can be seen that the project calendar schedule consists of a sequence of stages and a list of works. The project consists of four stages (49 working days): preliminary work (9 days); project preparation (8 days); project implementation (27 days); completion of the project (5 days). All stages are performed sequentially after the completion of the previous ones, which is associated with the specifics of the subtasks (work) performed. The duration of the work is indicated taking into account the expert estimates of the company's specialists. Some tasks are planned to be completed within two working days instead of one, due to possible delays due to reasons beyond the control of employees (power outages, emergency meeting, etc.) During the course of the project, the duration of such works can be reduced to one day and the total duration of the project will be reduced. In addition, the key milestone of the project on 09/07/2020 is shown, which means that by the specified date all stages and work must be completed, all documents developed and approved.

Table 3 provides a project resource sheet listing all resources used and estimated costs. For the project executors, approximate labor costs for the implementation of the project in connection with trade secrets are indicated.

It is shown that for participation in the project, one staff unit of labor resources is involved, with the exception of the account manager - two staff units. Expenses are calculated taking into account the fact that the project will be implemented without interrupting the main work, but taking into account the time for performing the main duties (2 h a day). Labor resources are not planned to be attracted overtime. Also, the resource list contains the standard rate (price) for the purchased software. As part of the project for the implementation of a customer interaction system, it is recommended to use the services of Bitrix24 on a turnkey basis, as well as to train four employees of the company to further work with the system. The total costs are shown for the purchased material resources and one-time costs. The total project budget is 360,000 rubles.

Thus, in the proposed project for automating interaction with the company's clients based on Bitrix24, the main components have been developed for its feasibility study. A recommended distribution of responsibilities for all project participants has been prepared. The schedule is drawn up taking into account the deadlines for completing subtasks recommended by the company's specialists, which can be clarified upon the fact of making a decision to start the implementation of the project. When implementing the project, it is necessary to make adjustments to the project estimate. However, the proposed project does not cover all emerging issues and requires the development of a project risk management plan and a preliminary assessment of the economic efficiency of the implementation of the project to improve the business processes of interaction with the company's clients based on CRM Bitrix24 in order to determine its investment attractiveness.

**Table 2.** List of project tasks (compiled by the authors based on research materials).

| Task name | Duration | Start | The ending |
|---|---|---|---|
| Project | 49 days | Wed 01.07.20 | Mon 07.09.20 |
| 1. Preliminary work | 9 days | Wed 01.07.20 | Mon 13.07.20 |
| Creation of an implementation group | 2 days | Wed 01.07.20 | Th 02.07.20 |
| Clarification of information on hardware and software | 3 days | Fri 03.07.20 | W 07.07.20 |
| Formation of ideas about the project, the approaches used, goals, tasks | 3 days | Fri 03.07.20 | W 07.07.20 |
| Clarification of requirements for implementation results | 3 days | Fri 03.07.20 | W 07.07.20 |
| Estimation of terms, resources, types and volumes of work | 2 days | Wed 08.07.20 | Th 09.07.20 |
| Assessment of the financial capabilities of the company | 2 days | Fri 10.07.20 | Mon 13.07.20 |
| 2. Project preparation | 8 days | W 14.07.20 | Th 23.07.20 |
| Distribution of powers and responsibilities | 2 days | W 14.07.20 | Wed 15.07.20 |
| Determination of organizational and technical requirements | 3 days | Th 16.07.20 | Mon 20.07.20 |
| Working group briefing | 1 day | W 21.07.20 | W 21.07.20 |
| Analysis of the main areas of costs | 2 days | Wed 22.07.20 | Th 23.07.20 |
| 3. Implementation of the project | 27 days | Fri 24.07.20 | Mon 31.08.20 |
| Clarification of the list of required software | 2 days | Fri 24.07.20 | Mon 27.07.20 |
| Acquisition software | 7 days | W 28.07.20 | Wed 05.08.20 |
| Installing and configuring software | 15 days | Th 06.08.20 | Wed 26.08.20 |
| User training | 3 days | Th 27.08.20 | Mon 31.08.20 |
| 4. Completion of the project | 5 days | W 01.09.20 | Mon 07.09.20 |
| Software testing | 3 days | W 01.09.20 | Th 03.09.20 |
| Commissioning | 3 days | W 01.09.20 | Th 03.09.20 |
| Preparation and signing of an act of completion of work | 2 days | Fri 04.09.20 | Mon 07.09.20 |
| Completion of the project | 0 days | Mon 07.09.20 | Mon 07.09.20 |

**Table 3.** Project resource sheet (compiled by the authors based on research materials).

| Resource name | Maximum units | Standard rate, rub./month | Labor costs | Costs, rub |
|---|---|---|---|---|
| Head of SRD | 100% | 20 000,00 | 294 h | 36 750,00 |
| Engineer | 100% | 15 000,00 | 78 h | 7 312,50 |
| 1C Specialist | 100% | 20 000,00 | 186 h | 23 250,00 |
| Testing Engineer | 100% | 15 000,00 | 156 h | 14 625,00 |
| Head of MPSD | 100% | 20 000,00 | 48 h | 6 000,00 |
| Client Relationship Manager | 200% | 15 000,00 | 18 h | 1 687,50 |
| General Director | 100% | 30 000,00 | 42 h | 7 875,00 |
| Executive Director | 100% | 25 000,00 | 42 h | 6 562,50 |
| Bitrix24 | 1 PC | 144 000,00 | | 144 000,00 |
| Training | 4 persons | | | 60 00,00 |
| Bitrix24 turnkey | 1 service | | | 50 000,00 |
| Total amount | | | | 358 062,50 |

## 5 Conclusion

The current economic situation in the country determines the development of trade. Most of the market is occupied by chain stores and online stores. Companies that develop these particular areas hold strong leading positions among competitors. However, only those companies that use competent management using modern information technologies are able to maintain such positions. However, only those companies that use competent management using modern information technologies are able to maintain such positions. An important element of management is the automation and improvement of the company's business processes. Business processes of interaction with clients are of particular importance. The presence of a well-developed system of business processes, the timely and most complete automation of these processes makes the company more attractive not only for clients, but also for investors.

The project for the implementation of a system of interaction with clients of the investigated company based on CRM Bitrix24 includes a stage of preliminary work, as well as stages of preparation, implementation and completion of the project. The main tasks of the project are carried out at the implementation stage and involve the purchase of the Bitrix24 CRM system, its installation, configuration and user training. As a result of the implementation of the project, it is expected to increase the productivity of employees; ensuring effective and transparent interaction with clients; quality control of work with clients, analysis of satisfaction; control of the return of financial documents, prevention of problems. The project provides only for the purchase, installation and configuration of software, as well as training for employees. As a result of the study, a description of the rationale and content of the project was prepared, a preliminary calendar plan and network schedule were developed, and a preliminary assessment of the project budget was carried out.

Thus, the implementation of CRM Bitrix24 will become the basis for increasing the company's client base, which contributes to an increase in profits and market value for investors. The implementation of the proposed project will improve the studied business process using modern network technologies, which is in line with the long-term strategic goals of the company. In addition, the transparency and scalability of the company's activities will increase, and its position among competitors will strengthen.

## References

1. Abbad Andaloussi, A., Burattin, A., Kindler, E., Weber, B., Slaats, T.: On the declarative paradigm in hybrid business process representations: a conceptual framework and a systematic literature study. Inf. Syst. **91**, 101505 (2020)
2. Assis Neto, F.R., Santos, C.A.S.: Understanding crowdsourcing projects: a systematic review of tendencies, workflow, and quality management. Inf. Process. Manage. **54**(4), 490–506 (2018)
3. Baiyere, A., Salmela, H., Tapanainen, T.: Digital transformation and the new logics of business process management. Eur. J. Inf. Syst. **29**(3), 238–259 (2020)
4. Bogdanov, D., Sazonov, S.: CRM systems. In: Sales Efficiency Improvement Tool. NovaInfo.Ru, vol. 115, pp. 46–48 (2020)
5. Bosikova, D., Volik, M. Possibilities of business management using the BITRIX 24 service. In: Young Scientists in Solving Urgent Problems of Science: Proceedings of the IX International Scientific and Practical Conference, pp. 20–24 (2019)
6. Dmitrieva, A., Kopysheva, T.: Development of an automated information system for an interior design studio. In: Actual Problems of Mathematical and Technical Sciences, Cheboksary, pp. 226–229 (2018)
7. Fabian, B., Bender, B., Hesseldieck, B., Haupt, J., Lessmann, S.: Enterprise-grade protection against e-mail tracking. Inf. Syst. **97**, 101702 (2021)
8. Gomes dos Santos, F., Machado, L., Pinheiro, R., Paes, A., Braganholo, V.: Querying XML documents using prolog engines: when is this a good idea? Inf. Process. Manage. **56**(5), 1753–1770 (2019)
9. Guerola-Navarro, V., Gil-Gomez, H., Oltra-Badenes, R., Sendra-García, J.: Customer relationship management and its impact on innovation: a literature review. J. Bus. Res. **129**, 83–87 (2021)
10. Ibragimova, R., Yakovleva, A.: CRM: accumulation of company information flows in order to effectively manage customer behavior. Bull. Ivanovo State Univ. Ser. Econ. **3**(41), 64–67 (2019)
11. Ishmuradova, I., Makhmutov, I., Shakirova, A.: Development and improvement of business process in the system «Bitrix 24» with the use of web technologies. Int. J. Eng. Technol. (UAE) **7**(3.27 Special Issue 27), 550–555 (2018)
12. Koch, H., Curry, P., Yan, J., Zhang, S., Milic, N.: How consumer technology is changing the it function: a multi-case study of three Fortune 500 companies. Inf. Syst. Manag. **36**(4), 336–349 (2019)
13. Kuchukov, R.: CRM as a key element of the management system of a modern company. Russ. Econ. Internet J. **3**, 48 (2019)
14. Litvinenko, I., Gurieva, L., Baburina, O., Ugryumova, M., Kataeva (Sventa Yarvik), V.: Tendencies and features of innovation management in the activities of businesses. Int. Bus. Manage. **10**(22), 5397–5405 (2016)

15. Magomedova, F.: Implementation of CRM systems to improve business competitiveness. Econ. Entrepreneurship **4**(117), 778–781 (2020)
16. Mantsivoda, A., Ponomaryov, D.: Towards semantic document modelling of business processes. Bull. Irkutsk State Univ. Ser. Math. **29**, 52–67 (2019)
17. Nam, D., Lee, J., Lee, H.: Business analytics use in CRM: a nomological net from IT competence to CRM performance. Int. J. Inf. Manage. **45**, 233–245 (2019)
18. Poddubnaya, M., Kravchenko, N.: Implementation of the BITRIX24 CRM system and its integration with online platforms. Sci. Soc. Econ. Law **4**, 140–144 (2019)
19. Russell-Rose, T., Chamberlain, J., Azzopardi, L.: Information retrieval in the workplace: a comparison of professional search practices. Inf. Process. Manage. **54**(6), 1042–1057 (2018)
20. Sarkum, S., Syamsuri, A.R.: The role of marketing function for competitive advantage. Qual. Access Success **22**(180), 32–39 (2021)
21. Slanova, A., Volik, M.: Features of the analysis of the company's business processes to improve the efficiency of customer service. Econ. Manage. Probl. Solutions **6**(1), 84–89 (2019)
22. Storm, C.P., Scheepers, C.B.: The impact of perceived work complexity and shared leadership on team performance of it employees of South African firms. Inf. Syst. Manag. **36**(3), 195–211 (2019)
23. Suoniemi, S., Terho, H., Zablah, A., Olkkonen, R., Straub, D.W.: The impact of firm-level and project-level it capabilities on CRM system quality and organizational productivity. J. Bus. Res. **127**, 108–122 (2021)
24. Suoniemi, S., Meyer-Waarden, L., Munzel, A., Zablah, A.R., Straub, D.: Big data and firm performance: the roles of market-directed capabilities and business strategy. Inf. Manage. **57**(7), 103365 (2020)
25. Xu, X., Qian, H., Ge, C., Lin, Z.: Industry classification with online resume big data: a design science approach. Inf. Manage. **57**(5), 103182 (2020)
26. Volik, M., Kopysheva, T., Mitrofanova, T.: Methodological features of the implementation of corporate information systems for managing a manufacturing enterprise. In: Young Scientists in Solving Urgent Problems of Science: Proceedings of the IX International Scientific and Practical Conference, pp. 366–369 (2019)
27. Volik, M., Kovaleva, M.: Features of automation of business processes of interaction with customers. In: SPBPU IDE 2020: Proceedings of the 2nd International Scientific Conference on Innovations in Digital Economy: SPBPU IDE-2020, vol. 63, pp. 1–6, October 2020
28. Vătămănescu, E., Pînzaru, F., Zbuchea, A., Andrei, A.G., Nicolescu, L.: The influence of competitiveness on SMES internationalization effectiveness. online versus offline business networking. Inf. Syst. Manage. **34**(3), 205–219 (2017)

# Opportunities and Barriers to Using Big Data Technologies in the Metallurgical Industry

Tatiana Verevka[1]([⊠]) [iD], Andrei Mirolyubov[1] [iD], and Juho Makio[2] [iD]

[1] Peter the Great St. Petersburg Polytechnic University, Saint Petersburg, Russia
verevkatv@mail.ru
[2] Leer University of Applied Science, 26723 Emden, Germany
juho.maekioe@hs-emden-leer.de

**Abstract.** One of the key areas of digital transformation of modern metallurgical production, which involves a high degree of automation as well as the use of complex technological systems, is the adaptation of production and business processes to new IT technologies for collecting and processing information. The purpose of this work is to study the conditions and business prospects for using big data technologies to improve the efficiency of production and operational activities of metallurgical enterprises in the context of digital transformation. The article examines the problems and prospects of using big data analysis tools to obtain significant economic results in an enterprise, considers the main barriers to the introduction of big data technologies in the industry and ways to overcome them, and analyses the results of the implementation of technologies for the construction of big data platforms of the data lake class at metallurgical enterprises in Russia. The results of the study show that the use of machine learning technologies and predictive analytics tools based on big data platforms could have a significant impact on reducing operating costs, increasing labour productivity, and improving the efficiency of metallurgical production.

**Keywords:** Big data · Internet of things · Machine learning · Big data application platform · Metallurgical industry · Sensor network · Predictive analytics

## 1 Introduction

Big data analysis technologies have become particularly relevant in the last few years, and their use by enterprises is growing exponentially. According to International Data Corporation (IDC) forecasts, the global market for big data processing and business intelligence technologies will grow by an average of 13.2% per year in the coming years and will reach $274.3 billion by 2022. The main driver for investment in big data and analytics is large enterprises with 1,000 or more employees, which account for two-thirds of the total market volume. The industries that currently collectively account for more than half of the world's investment in these technologies are banking, discrete manufacturing, professional services, continuous manufacturing, and federal governments. The leading segment is industrial enterprises, which account for 18% of the

© Springer Nature Switzerland AG 2021
D. Rodionov et al. (Eds.): SPBPU IDE 2020, CCIS 1445, pp. 86–102, 2021.
https://doi.org/10.1007/978-3-030-84845-3_6

big data consumer market [1]. In contrast to the IDC, other analysts give less optimistic forecasts of market growth. MarketsandMarkets experts believe that the volume of the global big data market will grow from \$138.9 billion in 2020 to \$229.4 billion by 2025, with a compound annual growth rate (CAGR) of only 10.6% over the forecast period [2]. This is confirmed by a survey conducted by Gartner of 199 members of the online community of IT leaders and business leaders, which showed that while some companies are investing in big data, many remain at the experimental stage. The reason for this situation, which can lead to a slowdown in investment, according to Gartner, is that many companies prefer to invest in other IT initiatives to build their infrastructure, citing the fact that the immediate return on investment in big data is difficult to calculate, and many big data projects do not have a tangible return on investment (ROI) that can be determined in advance.

In the modern literature, studies on the use of big data technologies in various spheres of economic activity are well represented: financial [3], transport [4], hospitality [5], public services [6], and so on. In-depth studies have been conducted on the impact of the digital economy and big data technologies on the production processes of industrial enterprises. These include the digital transformation of enterprises, which makes it possible to create a continuous production process through the synchronization of business processes and the integration of all elements of control, modelling, and enterprise management [7–9]. However, the range of these issues has not yet been fully investigated in relation to the enterprises of the metallurgical industry. Meanwhile, modern metallurgy is a branch of the economy with a high degree of automation of production, where complex technological systems are used. One of the key directions of the digital transformation of metallurgical production is complex work with data. Therefore, the purpose of this work is to study the conditions and business prospects for using big data to improve the efficiency of production and operational activities of enterprises in the metallurgical industry in the context of digital transformation. The study process uses statistical data from Strategy& and Statista; reports from the Russian Big Data Association and the Center for Strategic Research; data from annual ratings (Global Innovation 1000, Annual Performance 2500 world's top R&D investors, etc.); forecasts and estimates of analysts from companies such as IDC, PwC, BCG, Cisco, New Vantage Partners, McKinsey, Gartner, MarketsandMarkets, and Forrester; and analytical reports of IT companies (SAP SE, Oracle, Cisco Systems, Dell Technologies, T-Systems) as well as analytical data from leading Russian companies in the metallurgical industry.

## 2 Opportunities for the Development of Big Data Technologies in the Operational Activities of Metallurgical Enterprises

### 2.1 The Nature and Significance of Industrial Big Data

The nature of the data of a modern manufacturing enterprise falls into three general categories, namely: a) data that is generated by traditional methods (for example, data and reports of MES, ERP, or CRM systems); b) data generated during the operation of technological equipment in the process of performing a production task (readings of sensors, controllers, sensor equipment, etc. generated during the operation of APCS and UPC); and c) data from social media platforms [10]. The characteristics that relate to big data groups (b) and (c) are their volume, the speed at which they are generated, the variety of sources and content, and the economic value that the data can have. The most comprehensive definition of big data, proposed by Gartner [11], defines it as complex data sets that encompass a mixture of structured, semi-structured, and unstructured data that can be accounted for, stored electronically, and then analysed to establish patterns, trends, and associations. At the same time, in a practical sense, the term "big data" means not only the totality of data and the means of fixing it, but also a set of tools, approaches, and methods for processing it for subsequent use. Therefore, one of the most important subsections of big data is the means of machine learning (ML) and artificial intelligence (AI). ML is a method by which AI is able to independently learn from examples and mistakes, extracting patterns and making informed conclusions. ML tools show better results when working with large datasets, helping to reduce costs and optimize the use of time and resources. McKinsey experts believe that the use of big data will allow manufacturers to reduce the cost of product development and assembly by up to 50% and reduce working capital by up to 7% [8]. Moreover, at a higher level, thanks to the data sent by smart devices, manufacturers will be able to accurately determine the preferences of consumers and shape the characteristics of future products.

Of course, modern enterprises need new tools and procedures for analysing big data, because by combining them with traditionally collected data into a single cluster for analysis and forecasting purposes, an enterprise can stimulate production growth, ensure innovation, increase efficiency, and gain a competitive advantage in the market. In 2020, according to the Research and Markets report, 90% of business professionals and corporate analysts said that data and analytics were the keys to their organization's digital transformation initiatives [12]. Competition around new business models should be designed to enable data utilization to enhance industrial applications that are beneficial for end users [13]. In order for companies to remain efficient and competitive, they must adapt their business processes to new IT technologies for collecting and processing information, such as big data, the Internet of Things (IoT), cloud computing, and so on. In the metallurgical industry, big data and the IoT are two interrelated technologies that can optimize a company's entire supply chain, sales and marketing, and, most importantly, the production process [14].

## 2.2 Prospects for the Use of Big Data Technologies in the Operational Activities of Metallurgical Enterprises

The key questions applicable to the assessment of the prospects for the introduction of big data technologies in metallurgy are as follows: "What data models currently describe the production processes of a metallurgical enterprise?" and "On what methodological and technological basis will the processes of collecting, processing, and analysing big data be implemented?". Based on the information presented in the previous section, it should be noted that modern sensor technologies allow enterprises to analyse events and processes that need to be considered in the context of the application of big data technologies. For example, in the technological process of steel production, the data system includes information about the chemical composition of the alloy, the consumption of raw materials, the temperature of the furnace, the pressure of the rolls on the metal strip, cooling modes, and so on. Many systems and sensors are involved in ensuring the production process at metallurgical plants, which continuously record and transmit data to technological services. Using these data sources, it is possible to implement a model for optimizing the production cycle of an enterprise. At the same time, analytical applications of big data need to process a sufficiently large retrospective of data to draw conclusions about the real business value of the products produced [15]; applications for processing complex events (CEP—Complex Event Processing) are also used. With the introduction of IoT technologies, the sensors of the technological equipment of steel production will provide information about the parameters of the technological process in real time. Thus, the company lays the foundation for data collection and processing, but the tasks of analysing and predicting the performance of the production task remain unresolved. This is due to the following reasons: first, forecasting changes in the production process is difficult due to the fact that the available datasets may not be sufficient to cover the entire operational context; second, metallurgical companies produce products tailored to the specifics of a particular customer based on the formed order portfolio, so there are no standard procedures, as there are, for example, in mechanical engineering, which would facilitate the forecasting process. Thus, for a comprehensive solution to the problem of data management, an enterprise needs to create industrial big data analysis platforms (BDAPs).

The use of BDAPs for manufacturing enterprises has the following main advantages:

1. Improvement of an enterprise's production processes [16]
   Using industrial BDAPs, companies can process data generated in large volumes and at high speeds in real time and solve problems that arise in the production process to achieve their goals.
2. Optimization of deliveries and purchases
   With a set of IoT applications integrated into an enterprise supply chain, a manufacturing company has access to a huge amount of data on the execution of operations at all stages of the supply chain. By using the data obtained to carefully plan processes and make decisions, enterprises can minimize risks in the supply chain and avoid disruptions. This will also involve the relevant applications of the BDAP.
3. Customer support and warranty obligations
   Another use case for big data that has many advantages is the organization of consumer support and warranty obligations. Enterprises will be able to achieve significant savings in the resources for fulfilling warranty obligations if they use big data tools. Big data applications are able to detect violations of the quality and configuration of orders and eliminate them much faster than before.
4. Strategic and operational planning
   The implementation of the BDAP in an enterprise allows work assignment planners to improve the quality and optimize the output of products by assessing the impact of external events and adjusting performance strategies to take into account the changes that these events cause.
5. Financial management (analysis and forecasting)
   The financial forecasting capabilities of an enterprise can also be expanded through the introduction of big data technologies. They take into account many more factors and use all available historical data to develop and analyse the "what if" model of financial scenarios (sensitivity).

6. Saving income or avoiding losses

   There are three ways that big data technologies can help a manufacturing company maintain revenue or prevent losses. First, it is possible to analyse customer behaviour patterns for inconsistencies with expectations. Second, by analysing data related to changes in product quality, an enterprise can improve its operational efficiency. Third, by extracting information from external sources, such as social media, data scientists can make predictions about changes in customer behaviour, which are then taken into account when evaluating future profits.

### 2.3 Problems of and Potential for the Use of Industrial Big Data Platforms in the Conditions of a Metallurgical Enterprise

The platform that ensures the use of big data across the entire enterprise should provide the ability to work with all types of data in arbitrary formats and should include the following components [16, 17]:

- visualization and detection tools that allow the presentation of data to the user in a way that is convenient for perception, understanding of content, and searching in various sources;
- digital and text data analytics tools in batch and real-time mode;
- streaming data processing systems; and
- tools for supporting traditional data warehouses and online transaction processing systems.

There are six key capabilities that a BDAP must have in order for the enterprise using it to realize all its advantages [18, 19]. These include:

1) Data collection and aggregation

   As industrial companies generate more and more data, for example, from various devices and machines using the IoT, the BDAP that they will implement should make it possible to collect and aggregate data from various sources. Different types of data should be integrated and normalized using this platform, and their use should be consistent at different stages of the analysis. Finally, the platform should make it easier for companies to scale and use distributed file systems.

2) Advanced on-demand analytics

   Another key feature of BDAP is the customization of analytics. The needs and problems that a production company tries to solve with big data analysis are constantly changing and have different requirements, so the analytical procedures must be customizable to meet these needs [20]. The best way to do this would be to create libraries that contain all the different analytical procedures that are necessary in order to be able to test and change them quickly.

3) Flexible architecture, independent deployment

   A very important feature that makes it easier to scale BDAP is that it can provide high flexibility for deploying both in the local system and in the cloud. This means that the architecture must be flexible to allow new technologies to be deployed without interrupting the entire system.

4) Scalability and customizability

A BDAP should be designed in such a way that it provides a high degree of scalability. This capability will allow data scientists to use standardized application programming interfaces (APIs) and models and allow them to modify them to adapt to the possible acquisition of new devices and equipment that can generate new types of data for an enterprise. Ideally, the platform should also support third-party developers and the tools they create.

5) Harmonization

The BDAP that an enterprise is going to implement should be able to combine data, specialists, management tools, and analytics into a single whole so that all components can work both in isolation and in interaction with each other. To do this, the platform should be able to plan and manage data and machines, as well as how they interact with each other.

6) Modern user interface

Another necessary requirement that a BDAP must meet is the availability of a modern and intuitive user interface. Given the accumulated user experience, this will greatly facilitate the work and maintenance of such systems for employees.

However, the creation of new platforms of the BDAP class is complicated by the fact that, due to the growing volumes of data, the ideas about their accumulation and storage, which were given the name "data lake", are changing. Table 1 shows the distinctive features of data lakes compared to traditional data warehouses. A conventional data warehouse involves the distribution of input data streams and their reasonable placement; however, when data arrives in large quantities and in real time, they need to be collected in a certain way, and access problems must then be resolved. At the same time, the data warehouse stores mostly useful and in-demand things, where everything is merged into the "lake", including things that no one will ever need at all. In this sense, the metaphor "lake" is more informative, but for data lakes, adequate technologies are needed to extract the useful components. Among them, in particular, we can mention the pervasive data rush system, various implementations of CEP systems, search, data mining and text mining tools, and others.

Creating BDAP platforms requires systems that combine a variety of methods and tools for working with huge data flows and presenting them in a form that is accessible to a mid-level professional. Figure 1 shows the main technology providers using big data in the global market.

Although the market for ready-made BDAP solutions has not yet fully formed, nevertheless, we can talk about two main directions: one focuses on Hadoop/MapReduce and, according to the T-Systems survey, makes up 11% of the total market volume. The other focuses on in-memory data platforms, the most popular, which was chosen by 30% of the companies surveyed by T-Systems. The division into two directions is natural since they complement each other. Both technologies aim to adapt distributed hardware resources to work with large amounts of data; they can be compared to two approaches to high-performance computing, where there are cluster supercomputers with distributed memory and multiprocessor systems with shared memory. The similarity between clusters for high-performance computing (HPC) and Hadoop/MapReduce clusters is that the overall task must be broken down into fragments and ensure that they are executed in

**Table 1.** Comparative characteristics of traditional data warehouses and "data lakes".

| Comparative characteristics | Data warehouses | Data lakes |
|---|---|---|
| Data | Structured, processed | Structured/semi-structured/unstructured, raw |
| Processing | Schema-on-write | Schema-on-read |
| Storage | Expensive for large data volumes | Designed for low-cost storage |
| Agility | Less agile, fixed configuration | Highly agile; configure and reconfigure as needed |
| Security | Mature | Maturing |
| Users | Business professionals | Data scientists, etc. |

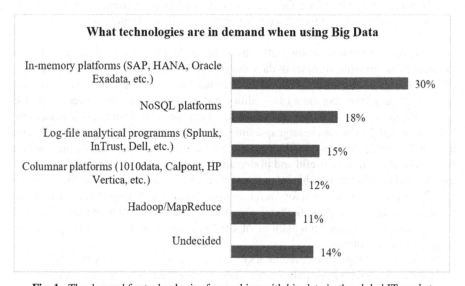

**Fig. 1.** The demand for technologies for working with big data in the global IT market

independent nodes, in order to then put everything together. On computers with shared memory and in-memory data grid configurations, such tasks can be solved. Hadoop technology is open-source software that is used for distributed storage and distributed processing of large datasets. The Hadoop ecosystem includes modules for data storage, data processing, data access, data management, security, and operations. Another powerful tool for managing large data sets is the MapReduce package, which provides a software model for filtering and sorting large amounts of data and then combining them into a single array.

Along with the advantages of using industrial big data platforms, it is worth noting the problems of their implementation in the conditions of a modern enterprise [21]:

*1) Cumbersome data and processes.*  Most manufacturing companies have pre-installed information systems and applications for analysing internal corporate information of the ERP or MES classes in their IT arsenal. These systems generate and store a large amount of diverse data, from marketing information to the results of the actual production process. Interacting with them often makes it difficult for contractors to implement big data platforms and integrate data for collaborative analysis using big data applications.

*2) Legacy and isolated systems.*  Outdated information systems and applications used by enterprises can also make it difficult to implement a big data platform since many of them initially did not have a big data access interface. Often, the IT infrastructure of enterprises is a set of disparate IT applications, which are built up with new components to meet the requirements of users. This leads to additional problems with integrating data from these systems into the shared storage.

*3) Hosting a big data platform.*  Another problem that complicates the process of creating a big data platform is the choice of a solution: whether to host the big data platform on the enterprise's own servers or whether it is advisable to use cloud technologies. Both options have their pros and cons, which will be different in each company, so it is necessary to conduct an analysis in order to make the optimal decision (taking into account all costs, security requirements, user access control, etc.).

*4) Poor project planning and implementation.*  The next problem that manufacturing companies face is when they try to upgrade or grow their ERP and MES systems, or when they try to combine them by creating a common data interface. Such decisions require very careful planning and preparation.

*5) Technology incompatibility.*  The general problem of creating an industrial big data platform and enterprise data warehouses is that companies initially used various IT tools to automate departments and processes (i.e. there was a "patchwork of automation"). This leads to a general decentralization of data, which will affect the implementation process of a big data platform.

*6) Lack of experience.*  Another common problem is the lack of specialists who are able to implement and maintain big data technologies in the manufacturing sector and ensure the effective use of analytics and forecasting tools.

*7) Inadequate tools.*  This problem lies in the fact that enterprises need additional tools for generating reports and tools for visualizing the results of analysis in order to provide access to them for specialists of other categories when making decisions.

## 2.4  Predictive Analytics Tools and Big Data Analysis Tools

As noted above, it is very important for manufacturing companies to identify factors that significantly affect changes in product quality and the occurrence of production failures, as well as optimize the cost of spare parts, service resources, and distribution, and this is where predictive analytics comes into play. Typical problems that predictive analytics solves in the manufacturing sector include:

- unexpected machine failures resulting in increased costs
- a lack of discoveries in datasets that lead to a decrease in the potential for using a big data platform
- low transparency in the supply chain
- the absence or late detection of defects in the products, which lead to additional costs

Predictive analytics is the process of using the results of data analysis to make predictions based on a data model. This process uses data along with analysis, statistics, and machine learning techniques to create a predictive model for predicting future events. This kind of analytics usually involves statistical models or machine learning models to make predictions. Typically, there are four main procedures that make up the predictive data analytics workflow (Fig. 2):

*1) Import data from all relevant sources.* In production, this will be data that is generated from all sensors built into the equipment, such as the operating time of the equipment, along with data from temperature and pressure sensors.

*2) Cleaning the data.* At this stage, data scientists must clean up the collected data by removing non-essential data or compensating for missing values using various methods, and then combine all the cleansed data together.

*3) Development of a predictive model.* In this stage, a predictive model is created. This model should be tested and trained on various sets of data collected to confirm that it has an acceptable level of accuracy. After the training phase is completed, the model can be tested on new datasets (for example, based on deviations from the expected machine performance data, a model that can try to predict when the machine will need repair or maintenance can be created).

*4) Integration of the model into the forecasting system.* The last step after the company is satisfied that the model is working properly is to integrate it into the system and the production environment.

To understand the challenges of big data analysis, it is necessary to break it down into separate stages or phases. Researchers [22–24] distinguish five consecutive phases. The first phase, "collection and recording", means that data analysts perform the process of identifying and storing large amounts of data, which will then need to be cleaned of errors for further processing. The second phase is information extraction and purification, where analysts, having sorted out all the necessary information, manipulate this data through processes such as manipulation, transformation, or purification to prepare it for analysis. Moving on to the third phase, the integration, aggregation, and presentation of data, analysts begin to diagnose the quality of the data collected in the previous stages. This is usually due to the fact that individual data may be missing or have extreme values that will negatively affect the analysis results. The first three phases relate to the data management process. The next two phases constitute the analysis itself. In the fourth phase, modelling and analysis, analytical applications begin to generate results through queries, data mining, and statistical analysis. After receiving the results, they are visualized. The last stage of the analysis procedure is the "interpretation" or "presentation" of the results. The results of the analysis, along with additional information, are passed on

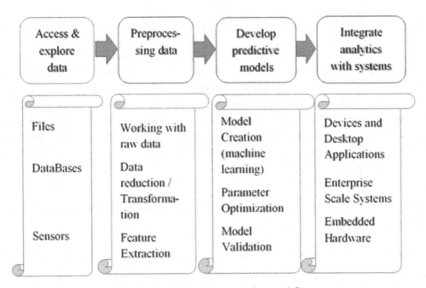

**Fig. 2.** Predictive data analytics workflow

to the next level managers, usually responsible for preparing the reports for the decision-making process, to begin processing them and presenting them to management. These specialists usually examine models using various assumptions and then search for patterns, after which they make decisions about the accuracy and reliability of the results obtained.

## 3 The Use of Big Data Technologies for Monetization and Achieving Economic Benefits in the Enterprises of the Russian Metallurgical Industry

Data monetization is the process that enterprises perform to generate revenue from their existing data sources in the process of discovering, collecting, analysing, storing, and interpreting this data. As mentioned above, the benefits of data analysis and its impact on the company's revenue are quite obvious, but it is advisable to look at it from the point of view of consumer-oriented companies that rely heavily on data analysis to maximize revenue. In recent years, these companies have had access to the vast amounts of data they collect, including network data, customer personal data, device data, location data, application data, and so on. By manipulating and analysing the data streams that their customers create, these companies can create unique customer profiles based on their individual needs. Thanks to the introduction of big data technologies, the traditional steel business can also change its face; for example, it can organize sales in a new way and become more focused on retail sales.

Specialists at the University of Texas in Austin [25] analysed datasets from Fortune 1000 companies for major industries in order to assess the impact that data analysis had on key performance indicators. The study showed that companies annually increase their revenues by more than $2 billion just by increasing the usability (access) of their data. In addition, the improved data quality resulted in an average increase of 16% in the companies' equity capital ratio. Thus, deep analysis of the production process, the use of machine learning technologies, and predictive analytics tools based on big data platforms can provide significant support in identifying sources of increasing production efficiency, reducing costs, and increasing labour productivity of metallurgical enterprises. Table 2 presents the results of the implementation of projects for the use of industrial big data and machine learning by enterprises in the Russian metallurgical industry, confirming the experience of using big data technologies in metallurgical enterprises of developed countries (Table 2).

**Table 2.** The results of projects using big data and machine learning by Russian enterprises in the metallurgical industry.

| Company name and year of implementation | Purpose of the project | Solution | Results |
|---|---|---|---|
| Magnitogorsk Metallurgical Combine 2016, http://www.mmk.ru | Construction of a mathematical model of steel melting in the oxygen converter shop | A model that optimizes the consumption of ferroalloys and alloying materials based on big data processing by machine learning tools on the Apache Hadoop platform was proposed | A service of recommendations for the staff of the steelmaking shop on optimizing the consumption of materials and minimizing costs has been introduced. Using the service allows the achievement of savings of 5% of the volume of ferroalloys at each production cycle, which equates to 275 million rubles per year |
| Novolipetsk Metallurgical Combine (NLMK) 2019, http://www.nlmk.ru/ | Creation of a system for predicting the chemical composition of the melt | Creation of a recommendation service (APM steelworker) based on machine learning algorithms, integrated with ERP and APCS | After the introduction of the recommendation service, the economic effect is 100 million rubles per year |

(*continued*)

**Table 2.** (*continued*)

| Company name and year of implementation | Purpose of the project | Solution | Results |
|---|---|---|---|
| Novolipetsk Metallurgical Combine (NLMK) 2019, http://www.nlmk.ru/ | Creation of a tool for storing and processing industrial data | Creation of a data lake based on ArenaData Hadoop | Data has been downloaded from 70 sources, and data maps of technological and production processes have been developed |
| Abinsk Electrometallurgical Plant (AEMZ) 2020, https://www.abinme tal.ru/ | Creation of a management system and end-to-end control of production assets based on end-to-end digital technologies | Creation of a single digital platform and digital model based on data from the APCS and the use of AI and machine learning, creating digital product passes | Labour productivity growth by 10% Energy efficiency growth by 20% Equipment downtime reduced by 15% |

# 4 Discussion

Despite the positive results of the introduction of big data technologies by leading metallurgical enterprises, it should be noted that their application in the Russian industry is currently of a utilitarian nature. The introduction of these technologies in Russia is hindered by outdated equipment, from which it is difficult to obtain information, and the unwillingness of the infrastructure to "read" and analyse data. According to the report by Digital McKinsey, "Digital Russia: new reality" (2017), the average age of power plants in the Russian metallurgy is 17 years [26]. The share of computer-controlled equipment (CNC) in Russia is only 10%, which is seven times lower than in Germany and nine times lower than in Japan. Despite the investments made in the renewal of the equipment fleet made over the past 15 years, Russian industrial enterprises demonstrate a low level of digitalization of processes. Even using modern equipment fitted with sensors, Russian manufacturers do not find any use for the data generated by this equipment. As a result, on average, only 1% of such data generated by sensors is used to improve the efficiency of an enterprise in Russia [26]. In addition, special attention should be paid to the Digital Transformation Index, which in Russia is currently 0.458, which corresponds to the level of lagging countries [27]. This fact indicates that the country has a low dynamic of the appearance of new products and services; new models of doing business are practically not used [28].

For the successful implementation of big data technologies in the metallurgy of developing economies, such as Russia, it is necessary to create specific conditions, overcoming a number of restrictions and barriers of technological, economical, and legal nature. The main restrictions include the following:

- A significant need for investments related to the modernization of production facilities and infrastructure [29, 30].
- Retraining and training of personnel capable of working with new technologies [31, 32].
- Insufficient (in comparison with developed economies) expenses for R&D in the sphere of digital technologies. Thus, according to the data of the Institute for Statistical Research and the Knowledge Economy (ISRKE) of the Higher School of Economics (HSE), the internal R&D expenses of Russian companies in the field of digital technologies in 2017 did not exceed 81.4 billion rubles (i.e. only 2.4% of total expenses on digitalization). This indicates serious backwardness and low competitiveness of domestic information technologies. Therefore, despite the high level of ICT usage in enterprises and the economic sector, their contribution to the total GDP of the Russian Federation remains consistently low: about 3% on average [28].
- Difficulties with the import of technologies and equipment as well as the lack of availability of information [33].
- Insufficient ICT infrastructure of enterprises and poor institutional environment. One of the problems of digital transformation is the general lack of funds for the digital modernization of manufacturing industries. At the same time, a paradoxical situation has developed—actual expenditure under the national project "Digital Economy" in 2019 amounted to less than 10% of the planned expenditure. According to the Accounts Chamber of the government of the Russian Federation, the main reason for insufficient spending is that the regional administrations have taken too long with agreements on subsidies, inter-budget transfers, and competitive procedures for state contracts. In addition, in most cases, there is no public expertise and detailed design and estimate documentation for the implementation of national project activities [34]. Therefore, the necessary conditions for the active development of digitalization nowadays are the creation of new institutions for innovation and mechanisms to stimulate the use of digital goods and services.

Evidently, these restrictions highlight significant risks for the successful implementation of national projects with the use of big data technologies in the metallurgy sector and their significant increase in price.

Currently, one of the most popular applications of big data in Russian industry is the monitoring of equipment conditions. By analysing information from equipment sensors, it is possible to detect even small deviations from the normal operating mode, quickly eliminate the causes of reduced performance, and prevent possible breakdowns. With predictive analytics, it is possible to move away from the traditional system of planned preventive repairs and reduce emergency stops and downtime. For example, as part of the project to automate the planning and management of repair work, implemented to reduce costly equipment downtime at the Vyksa Metallurgical Plant, the plant's equipment was equipped with NFC tags based on identification technology through wireless data transmission. As a result, the number of messages in the system about equipment downtime decreased by 50%, and the labour intensity of performing repair work decreased by 25%.

In our opinion, the creation of digital infrastructure and the implementation of big data analysis technologies in an enterprise should involve solving the following sequence of tasks:

1) Preparation of initial projects for creating a database. In this stage, machine learning models are built for individual processes or departments, the implementation of which can be both advisory and evaluative in nature. In the course of the work, recommendations about the value of the obtained machine learning models for the task and possible ways to increase the quality of the proposed models are formulated. The models are evaluated in terms of the potential economic effect and complexity of implementation for an enterprise. The order of priority of solving problems is also established. During the development of each project, the requirements for the future ecosystem of data and models are clarified, IT specialists and data analysts interact with production personnel, and the necessary training is provided.

2) Development of a data science strategy, which should include proposals for building a corporate data model; ensuring the model's life cycle, taking into account business value; and the creation of a "road map" for the implementation of an industrial big data platform that provides interaction with machine learning tools [35].

3) Implementation of a data lake that allows you to quickly process data using analytics and machine learning. As a result, it becomes possible to quickly accumulate the necessary data before setting specific tasks and reduces the time for testing and implementing machine learning models.

4) Accumulation of data in accordance with the needs of analytical tasks. The IT department of a company needs to organize the data storage space in advance and deal with its minimal structure, as well as link the data lake with the analytical ecosystem and ensure the information security requirements of the enterprise are met.

5) Create a layer of models and display them in a productive environment. Using the data lake and virtualization technologies, analysts deploy a layer of models and move on to creating them directly. As technologies and market requirements change, the composition of the data will change, which will probably require the modification or changing of models. In this case, tools that manage the life cycle of the model to select the most efficient model may be needed.

Thus, based on the analysis of the implementation of big data technologies in the enterprise environment, such solutions as CEP, machine learning, and the IoT allow production companies to create prerequisites for increasing efficiency. This means the ability to maximize profits by reducing the cost of production; facilitating and speeding up the work of specialists; increasing production safety; reducing expenditure on raw materials, waste, and equipment maintenance; and consistently preparing the conditions for the creation of autonomous production.

# 5 Conclusion

This article discusses the problems and prospects of using big data technologies by metallurgical enterprises to improve management efficiency. The importance and potential of using big data analytics in the metallurgical industry are shown. In the context of increasing global competition, the needs of enterprises for high-quality data processing are significantly increasing. In these conditions, big data technologies become a means of achieving competitive advantage by the enterprise and obtaining increased profits. Big data technology allows organizations to bring data from process control systems, transport logistics, and applied business systems together, significantly reducing resource consumption compared to traditional approaches. The achieved savings in resources and decision-making time indicate the need for the introduction of big data technologies as a promising tool for the development of the metallurgical business. Thanks to the use of machine learning and predictive analytics, enterprises are able to flexibly manage production and economic results.

# References

1. Worldwide Semiannual Big Data и Analytics Spending Guide, IDC. https://www.businessw ire.com/news/home/20190404005662/en/IDC-Forecasts-Revenues-for-Big-Data-and-Bus iness-Analytics-Solutions-Will-Reach-189.1-Billion-This-Year-with-Double-Digit-Annual-Growth-Through-2022.
2. Big Data Market by Component, Deployment Mode, Organization Size, Business Function, Industry Vertical and Region - Global Forecast to 2025, MarketsandMarkets. https://www.marketsandmarkets.com/Market-Reports/big-data-market-1068. Accessed 12 Mar 2021
3. Bataev, A.: Analysis of the application of big data technologies in the financial sphere. In: Proceedings of the 2018 International Conference "Quality Management, Transport and Information Security, Information Technologies", IT and QM and IS, pp. 568–572 (2018)
4. Ilin, I.V., Iliashenko, O.Y., Klimin, A.I., Makov, K.M.: Big data processing in Russian transport industry. In: Proceedings of the 31st International Business Information Management Association Conference, pp. 1967–1971. IBIMA, Milan, Italy (2018)
5. Verevka, T.: Development of industry 4.0 in the hotel and restaurant business. IBIMA Bus. Rev. (2019)
6. Furtatova, A., Kamenik, L.: Predictive analysis of water supply based on big data. In: Proceedings of the 33rd International Business Information Management Association Conference, pp. 8762–8767. IBIMA, Granada, Spain (2019)
7. Geissbauer, R., et al.: Industry 4.0 – Opportunities and challenges of the industrial internet assessment. PricewaterhouseCoopers (2014)
8. Yin, S., Kaynak, O.: Big data for modern industry: challenges and trends. In: Proceedings of the IEEE, vol. 103, no. 2, pp. 143–146. IEEE (2015)
9. Naumov, A., Bezobrazov, Y., Glushkov, S.: Perspective and efficiency of numerical analysis for advanced manufacturing technologies using big data. In: Proceedings of the 33rd International Business Information Management Association Conference, pp. 8539–8545. IBIMA, Granada, Spain (2019)
10. Alam, M., Muley, A., Kadaru, C., Joshi, A.: Oracle NoSQL Database: Real-Time Big Data Management for the Enterprise.1st edn. McGraw-Hill Education, New Delhi (2013)
11. Information Technology Glossary, Gartner. https://www.gartner.com/en/information-techno logy/glossary/big-data. Accessed 10 Mar 2021
12. Global Big Data Analytics Market Size: Market Share, Application Analysis, Regional Outlook, Growth Trends, Key Players, Competitive Strategies and Forecasts, 2019 To 2027, Research and Markets (2020)
13. Berawi, M.: Utilizing big data in industry 4.0: managing competitive advantages and business ethics. Int. J. Technol. 9(3), 430–433 (2018)

14. Subbaiah, K., Rao, K.: Value chain model for steel manufacturing sector: a case study. Int. J. Manag. Value Supply Chains **6**(4) (2015)
15. Jacobi, S., Krumeich, J.: Big data analytics for predictive manufacturing control – a case study from process industry. In: Chen, P., Jain, H. (eds.) Proceedings of the 3rd IEEE International Congress on Big Data. IEEE, Anchorage (2014)
16. The Case for an Industrial Big Data Platform: Laying the Groundwork for the New Industrial Age. General Electric (2013). https://www.ge.com/digital/sites/default/files/download_assets/the-case-for-an-industrial-big-data-platform-brochure.pdf. Accessed 14 Mar 2021
17. Yuhanna, N., Hopkins, B.: Big data is critical technology for insights-driven businesses. Forrester (2018)
18. Minelli, M., Chambers, M., Dhiraj, A.: Big Data, Big Analytics: Emerging Business Intelligence and Analytic Trends for Today's Business. John Wiley & Sons, Hoboken (2013)
19. Marjani, M., et al.: Big IoT data analytics: architecture, opportunities, and open research challenges. IEEE Access **5**, 5247–5261 (2017)
20. Mayer-Schönberger, V., Cukier, K.: Big Data: A Revolution That Will Transform How We Live, Work, and Think. John Murray, London (2013)
21. Big Data Adoption in Manufacturing: Forging an Edge in a Challenging Environment (2017). Protiviti. https://www.knowledgeleader.com/knowledgeleader/content.nsf/web+content/articlebigdataadoptioninmanufacturing. Accessed 12 Mar 2021
22. Kandel, S., Paepcke, A., Hellerstein, J., Heer, J.: Enterprise data analysis and visualization: an interview study. IEEE Trans. Visual Comput. Graphics **18**(12), 2917–2926 (2012)
23. Labrinidis, A., Jagadish, H.: Challenges and opportunities with big data. Proc. VLDB Endowment **5**(12) (2012)
24. Akerkar, R.: Advanced data analytics in business. In: Akerkar, R., et al. (eds.) Big Data Computing, pp. 373–400. CRC Press, Boca Raton (2013)
25. Barua, A., Mani, D., Mukherjee, R.: Measuring the business impacts of effective data. University of Texas at Austin (2017). https://www.datascienceassn.org/sites/default/files/Measuring%20Business%20Impacts%20of%20Effective%20Data%20I.pdf. Accessed 12 Mar 2021
26. Digital McKinsey: https://www.mckinsey.com/~/media/mckinsey/locations/europe%20and%20middle%20east/russia/our%20insights/digital%20russia/digital-russia-report.ashx. Accessed 12 Mar 2021
27. Osmanova, Z.: Monitoring the results of digital transformation in the Russian Federation on the basis of the national. Sci. Bull. Financ. Banking, Invest. **3**, 159–167 (2019)
28. Verevka, T., Gorbunov, E., Shpigalskiy, P.: Digital transformation of the Russian agricultural sector: technological and economic barriers. In: Proceedings of SPBPU IDE 2020: International Scientific Conference on Innovations in Digital Economy 2020, Article 26. ACM International Conference Proceeding Series, 3444497 (2020)
29. Rodionov, D., Rudskaia, I.: Problems of infrastructure development of 'Industry 4.0' in Russia on Sibur Experience. In: Proceedings of the 32nd International Business Information Management Association Conference, IBIMA 2018 - Vision 2020: Sustainable Economic Development and Application of Innovation Management from Regional Expansion to Global Growth, pp. 3534–3544 (2018)
30. Kudryavtseva, T., Skhvediani, A.: Effectiveness assessment of investments in robotic biological plant protection. Int. J. Technol. **11**(8), 1589–1597 (2020)
31. Branca, T., et al.: The challenge of digitalization in the steel sector. Metals **10**, 288 (2020)
32. Kudryavtseva, T., Skhvediani, A., Arteeva, V.: Theoretical analysis on the effect of digitalization on the labor market. In: European Conference on Knowledge Management, pp. 672–679 (2019)

33. Yong, W., et al.: Smart sensors from ground to cloud and web intelligence. IFAC PapersOnLine **51**(17), 31–38 (2018)
34. Sergeyev, S.: Digitalization of the system public administration and the budget process. Int. Sci. Pract. Forum (2019)
35. Cao, L.: Data Science Thinking: The Next Scientific, Technological and Economic Revolution. 1st edn. Springer, Cham (2018). https://doi.org/10.1007/978-3-319-95092-1

# End-to-End Digital Technologies
# in Industry

# Digitalisation for Improving Population Well-Being in the Arctic Area

Valeria Rakova[1] , Marina Bolsunovskaya[1] (✉) , Arseny Zorin[1] ,
Vladimir Fedorov[2] , and Yuliya Novikova[2]

[1] Peter the Great St Petersburg Polytechnic University, St Petersburg, Russia
`marina.bolsunovskaia@spbpu.com`
[2] North-West Public Health Research Centre, St Petersburg, Russia

**Abstract.** This study describes the development of information and technological support that accumulates information and analytical materials concerning the sanitary and epidemiological environments of the Russian Arctic population. Maintaining a healthy epidemiological situation is vital for the Arctic region. However, active development of the Arctic region is ongoing. The penetration of pathogens into the region is unacceptable.

Information and technological support for social and hygienic monitoring is being developed. Within the framework of the geoportal, a user's personal account is formed with the possibility of registering it. The required functionality of the portal has been developed; an interface has been implemented to improve decision-making efficiency based on the information provided to the user. Input data on the state of health and environmental factors of the Russian Arctic population allow us to analyse information on directions and regions. Data were collected from all the arctic zone regions and provided by the respective regions for the past year. The developed software allows processing input data and forming summary files in the following directions: 'Public Health', 'Human habitat', 'Socio-economic markers', 'Medico-demographic markers', 'Food safety', etc. The system creates summary files, which are further analysed by a separate software module with the possibility of visualisation by layers. The developed IT solution can significantly improve the sanitary and epidemiological well-being of the Russian Arctic population.

**Keywords:** Geoportal · Algorithm · Software · Information technology support · Russian Arctic

## 1 Introduction

During previous decades, the number of studies concerning the Arctic Region have increased. Different groups of scientists from the Nordic countries engage in ongoing research on the ecology, demography, economy, and industry of the Arctic. Among the most authoritative reports are the Arctic Human Development Report (AHDR) and Economy of the North (ECONOR).

© Springer Nature Switzerland AG 2021
D. Rodionov et al. (Eds.): SPBPU IDE 2020, CCIS 1445, pp. 105–117, 2021.
https://doi.org/10.1007/978-3-030-84845-3_7

The AHDR is prepared under the auspices of the Nordic Council of Ministers (the main forum for official Nordic co-operation, involving Denmark, Finland, Iceland, Norway, Sweden, the Faroe Islands, Greenland, and Åland) [38]. The first AHDR was an assessment of the state of human development in the Arctic; it was published in 2004 and initiated by the Stefansson Arctic Institute [1]. The second AHDR II was prepared 10 years later in 2014. The subtitle of the second issue was 'Regional Processes and Global Linkages', and it overviewed the critical and emerging challenges facing the Arctic region, including the impacts of global changes. The issues of sustainable development in the Arctic were discussed against the background of rapid changes. The AHDR II was conducted by an international group of leading scientific experts on Arctic issues, and the management and coordination of the project was executed by the Stefansson Arctic Institute [1].

The issues of ECONOR were prepared by the Sustainable Development Working Group of the Arctic Council (SDWG). The first phase of the ECONOR project was finished in 2006 (ECONOR I), and the second and third phases (ECONOR II and III) were prepared from 2007 to 2017 [7]. The ECONOR IV is currently being prepared, and it will provide an updated ECONOR report: 'The Economy of the North 2020' [8]. The objective of the ECONOR project is to comprehensively overview the economy of the circumpolar Arctic. The ECONOR reports contribute to harmonising socio-economic statistical data across national and regional borders in the Arctic region, improving circumpolar statistical and research knowledge base on policies for Arctic sustainable development. The project has formed the ECONOR scientific network: the ECONOR III project was conducted with Norway as the lead country, Canada, the United States, and the Saami Council as co-leads. The editorial group of ECONOR IV is based at Statistics Norway in close cooperation with CICERO and Université Laval, Quebec, and Canada. ECONOR IV also has close cooperation with the Canadian research network, Wealth of the Arctic Group of Experts (WAGE), and Business Index North (BIN) [8].

Moreover, regular reports on the Arctic issues were prepared by the international centre for research on regional development 'Nordregio' established by the Nordic Council of Ministers in 1997 [39]. Europe's scientific groups, entitled Barents Euro-Arctic Region (BEAR), also conduct Arctic research.

The mentioned reports covered most of the Arctic regions, but there are many studies devoted to specific regions, for example, studies dedicated to the Canadian Arctic [12, 14], Nordic Region [1, 13, 38], and Russian Arctic [10, 11, 20, 21]. Individual areas are also being studied. In the Russian Arctic, these are Chukotka, Yakutia [6, 10, 21], and others.

Studies in the field of the Arctic population's health are usually conducted by medical specialists. Different diseases were studied since they often occur differently in the Northern regions; the health of natives and those who came to the Arctic for professional activities were also investigated [13, 20]. For instance, some studies focused on diseases such as Hepatitis B and C, bacterial meningitis, tuberculosis, cancer, pneumococcal infections, anthrax [10, 12, 14, 19, 21, 22], and diseases carried by animals are separately investigated.

Many scientific groups are studying the impact of ecology on human health in the Arctic [1, 18, 39].

There has been an increasing interest among scientists and enterprises in the Russian Arctic in the past decades and an increase in its population. The State Programme 'Socio-economic Development of the Arctic Zone of the Russian Federation' has been in operation for years [42]. The International Arctic Forum 'Arctic: Territory of Dialogue', held unprecedentedly in 2010, is the largest Russian forum for discussing the Arctic region's development prospects [41].

However, the problem of automatic data collection for scientists' research remains. Only a minority of the mentioned studies reference the use of automation tools at the stage of data collection on the epidemiological or ecological environment of the Arctic region population.

The authors of this study have been working on creating a geographic information system for scientific research in the Russian Arctic since 2007, and this prototype of a geoinformation system has no analogues [11].

One of the key tasks for improving population well-being in the Arctic area is the scientific and methodological support and analysis of the state of sanitary and epidemiological well-being in the territory. An important issue is related to developing information and analytical software systems based on geographic information systems for monitoring the sanitary and epidemiological situation. This enables the conducting of a comprehensive analysis of the health status and environmental factors of the population in the Arctic zone of the Russian Federation (hereafter called the Russian Arctic) [26].

Given the need to create tools for the collection and visualisation of data on the health status and environmental factors of the Arctic population, the implementation of information and analytical systems for processing large volumes of social and hygienic data monitoring is relevant. This system can serve as a tool for the preparation of justified management decisions by public authorities to reduce the adverse impact of environmental factors on the population's health [26].

Notably, the Russian Arctic territory is allocated as an independent object of statistical observation. That is, the data are collected specifically for this object. The information and analytical system will allow the collection and analysis of data as efficiently as possible. This will make the preparation of annual issues of newsletters and materials on the state of sanitary and epidemiological well-being of the Arctic population possible, among other things.

To achieve this goal, the following tasks were performed:

- Comprehensive analysis and systematisation of data on the health status and environmental factors of the Russian Arctic population distributed by area and region
- Organisation of automated data collection
- Development of the architecture of the web system, and the formation of recommendations for changes and adaptation of the used technical, technological, and IT solutions
- Web system design development
- Software development, testing, and trial operation

## 2   Material and Methods

The input data of the health status and environmental factors of the population are a set of templates in the format of Excel workbooks, divided by directions and regions. Data for all the Russian Arctic regions were collected: Murmansk Region, Nenets Autonomous Area, Chukotka Autonomous Area, Yamalo-Nenets Autonomous Area, Komi Republic (partically), Sakha Republic (Yakutia), Krasnoyarsky Krai (partically), Arkhangelsk Region (partically), and the Republic of Karelia (partically). Data were provided by the respective regions for the past year.

Furthermore, the templates were processed by employees manually or using special software. Consequently, summary files were formed in various directions. For example, 'Water' for public health, the environment, radiological conditions, etc. [33].

Summary files were analysed by a separate software module, which converts the files into a special format for further sending to the server where the display layers were formed.

Various plugins are used for creating user roles and basic access rights differentiation; for detailed configuration of the rights of the created user category; for sending emails via the SMTP server; and for quick access to the settings of the selected SMTP server. Additionally, the plugins allowed the user to implement the ability to make informational mailings or point-to-point sending of emails to users through the administrative console as well as configure the site display for the visually impaired, the site's multilingualism, etc.

To implement the above-mentioned features and many other tasks, the development methodologies and tools were analysed and researched. Further, their choice was justified and implemented [27].

Ubuntu 20.04 Focal was chosen as the operating system (OS). This OS was freely distributed. This allowed the user to avoid additional costs for the purchase of licenses. Added to free distribution, this OS is an LTS version (long-term support). This guarantees long-term support—five years. Since the deployment of the entire portal does not require a graphical shell, it was decided to use the server version of the OS. The server version of the OS is distinguished because there is no graphical shell in it. This reduces the space occupied by the OS itself.

The main technology is Docker [35]. This technology allows the user to achieve the isolation of each service as if they were each running on a separate computer. Due to this partitioning, the user can avoid one service impacting another. And this applies both to the impact of the services themselves and their dependencies.

To remove the manual settings of all running images, special orchestration utilities were used. This allows the user to ensure the connection of all blocks to each other, and to ensure constant data storage. This allows the block to capture the previous data during the reboot, rather than starting a new installation.

The domain registrar is an important tool. The registrar allows the user to rent a domain name. The registrar is the entry point of the portal being developed. The Round-Robin technology was chosen to ensure the portal's fault tolerance. For the proper operation of the portal in this case, the services must be completely identical. The Round-Robin technology allows one domain name to set several addresses that will change constantly (the domain name will point to one address, then to another). If one of the addresses becomes unavailable, this technology will point to the available address, and the user will not notice the portal failure. Since both servers are completely identical, only one will be considered in the future [37].

This software helps the user determine how to manage an incoming request. It was planned to use Caddy as a web server. Caddy is a fairly lightweight server and is easily configurable. With it, all incoming requests for a registered domain name will be sent to the deployed site.

Since all software was deployed inside containers, they were inaccessible from the machine on which these containers were deployed by default. To provide access to this software, the user needs to configure the transition. This setting was made only for the site, since access to the other services from the outside was not needed. For the site, an additional database (DB) was deployed.

Since the site is the main entry point to portal services, most of the interactions were organised through it. Thus, the files uploaded through the personal account were saved to file storage due to the site's direct interaction with it. The storage is S3 compatible (Simple Storage Service). This is a widely-used format for organising data storage. This solution avoids direct interaction with the file system of the server on which all the services were deployed. This technology increases the entire port's security.

For the site to interact with the information system, the Application Programme Interface (API) should be added. The presence of this interface simplifies the interaction of the programmes with each other.

The Geographic Information System allows the user to interact with maps and geographic information. This software is proprietary and has a closed license. The design problem involves discovering new technical solutions and methods for the correct installation of the Geographic Information System. In the system, a list of layer sets in different directions was formed.

Including the characteristics of water resources and drinking water, atmosphere, and soil; medical and demographic indicators; socio-economic state of the territory; characteristics and specifics of food products; and indicators of resource pollution. The sanitary and epidemiological situation at water bodies, leading sources of pollution, food safety, etc. have also been evaluated [16].

To determine each layer's parameters, information was collected in the form of a questionnaire. The software module automatically loads the information into the system.

The information in the system is accumulated in various sections. These are 'Public health', 'Human habitat', 'Socio-economic markers', 'Medico-demographic markers', 'Food safety', 'Radiology safety', etc. The system methodology used allows for the accumulation and structuring of information.

The area of 'Public health' includes blocks related to medical and demographic indicators, morbidity of children and adolescents, school children, disability, socio-economic indicators of the territory, and some other indicators [16].

The 'Radiology safety' section includes blocks related to monitoring (by point and by time), the density of soil contamination with man-made radionuclides, the number of inhabitants in an area with different densities of soil contamination, and other indicators.

In the 'Food safety' section, special attention is given to the content of contaminants in food raw materials and food products.

The 'Human habitat' section is also considered: the atmosphere, water, and soil.

The system analysis resulted in a detailed structure and composition of the directions being determined. It was revealed that templates with aggregated data on the subjects were provided for some areas. Files can contain only a part of the data if the collection of information in the region is unavailable for objective reasons (for example, there are no control points).

The system implements the organisation of automated data collection. The data files will be uploaded to Geoportal by users in Excel workbook format.

The automated data collection module performs the following tasks:

– Collection of all saved files (templates)
– Fixing incorrect values in files
– Generation of a file for ArcGIS
– Formation of summary tables in the following sections: 'Food safety' and 'Human habitat'
– Formation of reporting forms

Invalid values that can be analysed by the module:

– Zero values in the coordinates of the object location
– Missing values
– Lowercase values in the cells where the numbers should be
– File structure violation [28–30].

After further analysis, a list of incorrect values can be supplemented.

The algorithm for automated data collection for the formation of pivot tables includes the following:

1. Creating a copy of the summary file
2. Creating a list of suitable input files
3. Select the input file from the list
4. Forming a list of analysed sheets
5. Reading and processing data from a single file
6. Writing to the summary file
7. If the list of input files is not complete, then go back to select the input file from the list
8. Saving the summary file

Based on this algorithm, a software module for automated documentation assembly has been developed. The software module collects data on the 'Habitat' direction from all regions in a single summary file.

The programme module includes:

1. Data processing module
2. File reader module
3. Data recording module
4. Basic module

5. GUI module
6. The 'Human habitat' data processing module, which includes the atmosphere, soil, and water submodules [17, 32].

Thus, the main methodology for constructing the directions' structure is a systematic approach that allows the user to structure and form data sets by direction and layer.

## 3 Results

The main result of this study is the developed information system (Fig. 1). The main component of the entire system is the ArcGIS Server. This server handles processing all data, uploading them, and interacting with the GIS system. All necessary configurations were also formed [28–30].

To allow the site to interact with the ArcGIS software and the data processing software with the system under development, three more components need to be implemented. These are the ArcGIS Portal, Web Adaptor, and Data Store.

The ArcGIS Portal is a component of the ArcGIS system. The portal provides shared access to maps, scenes, apps, and other geographic information. This component combines all geographical information and makes it available within the entire portal.

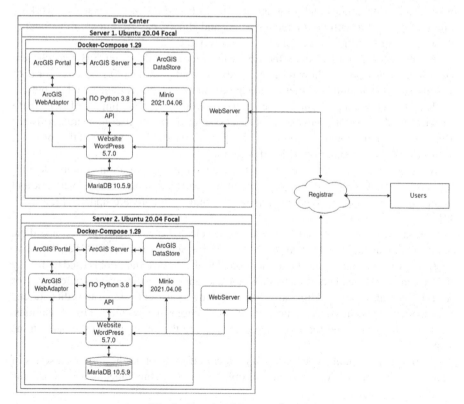

**Fig. 1.** Portal architecture diagram

This component allows the user to create, save, and publish web maps and scenes. Furthermore, it is possible to create and host cartographic web applications on the portal. The component also allows the user to search for GIS resources in the organisation and create groups for sharing GIS information with colleagues as well as provide links to GIS applications and publish maps and layer packages.

Additionally, the portal allows the user to access the ArcGIS API [28–30].

In the system architecture, the user can also select an Adapter that runs on an existing website and redirects requests to the ArcGIS Server [33–35].

The ArcGIS Data Store application is also implemented, which allows the user to configure the data store for the server used with the ArcGIS Portal. A relational storage is configured to solve the tasks set in the system.

A Docker-compose file was written to deploy the entire ArcGIS suite. It specifies the components of the failure—or unexpected behaviour—and where to store the data [17, 23, 32].

To create a Geoportal, various platforms were considered. As a result of the analysis, WordPress CMS was chosen for the implementation of the web resource.

The chosen platform has certain advantages. Notably, the ability to change the appearance of resources without changing the server part of the system, convenient administration of the resource and content, and high-quality security. The system has additional functionality for optimising the site to meet the requirements of search engines.

An important advantage of the platform is the availability of a considerable number of ready-made templates and plugins for solving trivial tasks. For example, working with tables, text formatting, interactive forms and sliders, etc. Additional functionality allows the user to easily implement new plugins in the CMS.

As the servers' main system, the OS of the Linux family, CentOS 8, was chosen. This choice was made to allow for a more flexible configuration of the system, for both security and automation. Another advantage of Linux systems is their free distribution.

The Docker containerisation system is required to reduce the impact of third-party libraries and the developer environment. This system allows the user to operate software in an isolated environment—containers. This ensures the stable operation of the software, regardless of the computer on which it is running.

Since several containers are required, the container orchestration system, Docker-compose, is used. This system allows the user to correctly configure the network and communication between containers, as well as automate software updates [2, 17, 23, 32].

Various tools of the ArcGIS software were selected for processing, mapping data, and providing access to maps: ArcMap, ArcGIS for Server, and ArcGIS Web Adapter.

Here, we characterise the above-mentioned tools, the ArcMap tool is used to display and explore sets of geodata. With this tool, the user can set symbols and prepare a map for printing and publishing. Additionally, ArcMap allows the user to create and edit datasets. This application is among the main desktop applications that has unique functionalities. These functions are used for various tasks, including the ability to analyse data on the map.

Also, it is important to publish geographical information in the form of a series of map services. The services provide multiple users with simultaneous access and the ability to work with data hosted on server computers. The user can also include this data in their GIS products on the web or mobile devices.

Thus, ArcMap is necessary in building a map and drawing layers of data on it, and ArcGIS Server stores this data as services and allows the user to provide access to them.

ArcGIS Server, a component of server software, plays an important role. It makes geographical information available to any Internet user, if necessary. This is achieved through the use of web services. These services allow the server to receive and process requests for information sent by other devices [3]. The ArcGIS Web Adapter links the ArcGIS Server and the site where the maps will be selected. The Adapter runs on an existing website and redirects requests to ArcGIS Server computers. This allows the user to create a single point that processes and directs incoming requests to the Server on the site [28–30, 36].

As described above, ArcGIS Server was used to provide access to maps and data. ArcGIS Online stores maps directly in the cloud storage. ArcGIS Server allows the user to keep maps and databases on your server and fully control their performance and access to them. This is an important security advantage.

The chosen software for website development, WordPress, requires the creation of its own database. The choice of a special technology induces an increase in the speed of error correction and an increase in the level of security provided.

The efficiency of ArcGIS software, like WordPress, requires a personal database [33–35].

To organise the storage of files uploaded by the user, it was decided to use S3 compatible storage, Minio [37]. This storage is lightweight and practically does not increase the number of resources consumed.

Round-robin technology will ensure the portal's fault tolerance. If one of the servers is unavailable, Round-robin will allow the user to use the other with almost no downtime. This technology was chosen because it allows the user to organise fault tolerance without an additional router and computing cluster. Additionally, this technology requires only a DNS record from the domain name registrar to be configured [24].

The site layouts are designed under the following requirements:

- Home page, including news about sanitary and epidemiological well-being in the Russian Arctic, brief information about the Geoportal, and the possibility of switching to the Geoportal [4]
- Documents, including laws, regulations, and development strategies
- Analytical materials (for example, publications on health and sanitary-epidemiological well-being, as well as collections of conference articles), news, and events.

In this study, six variants of the first screen of the main page were developed, and the original version of the colour scheme was aligned with the approved logo. In addition to the pages corresponding to the above structure, layouts for internal pages of news, analytical materials, and events, as well as an authorisation screen, have been developed [27].

# 4  Conclusion

One of the main strategic state tasks set by the Government of the Russian Federation for the Sustainable Development of the Arctic is:

- Improving the effectiveness of measures to preserve and promote the health of the indigenous and migrant populations.
- Minimisation of the negative anthropogenic impact on the Arctic environment. This impact is due to the current economic and other activities and the implementation of investment projects.
- Study of the impact of harmful environmental factors on the Arctic population's health and the scientific justification of a set of measures aimed at improving the habitat of the population and preventing diseases.

The active development of the region is associated with a significant and intensive attraction of human resources to work in various industries and in the social sphere. This is due to the many difficulties caused by the impact of the North's specific climate. In this case, the primary task is to preserve the health of the people, as the main condition for the planned sustainable development of this region's economy in the difficult climatic conditions of the Arctic.

It is necessary to consider the conditions and factors that have significant impact. This is also the influence of low temperatures and permafrost conditions on the organisation of public services in populated areas. Drinking water supply is characterised by low water mineralisation. This induces an endemic incidence of diseases of the endocrine, bone, and cardiovascular systems. Also, there is a high risk of accidents of water-wire and sewer systems in permafrost conditions. This threatens the population's epidemic safety.

Based on the analysis, it can be concluded that the Arctic territories belong to the zone of uncomfortable areas with elements of pronounced extremity. This can be seen from the analysis of the totality of climatic characteristics, considering the actions of natural, anthropogenic, and social factors and their combination.

Therefore, for the organisation of sustainable economic development in the Arctic zone, it is necessary to organise and implement comprehensive measures to prevent the morbidity of the population and create conditions for ensuring sanitary and epidemiological well-being and minimising health risks [25, 37].

The development of a software package using geoinformation technologies to ensure sanitary and epidemiological well-being in the Russian Arctic territories allows us to solve many problems. Based on the information analysis, management decisions are developed to reduce the impact of anthropogenic and non-anthropogenic factors on the population's health.

Thus, in the project for the development of the information complex, the following scope of work was completed:

- A comprehensive analysis and systemisation of data on the state of health and environmental factors of the Russian Arctic population, distributed by direction and region, was conducted;

- The organisation of automated data collection and the formation of a set of reporting documentation was conducted.
- The development of the web-system architecture and the formation of recommendations for changes and adaptation of the used technical, technological, and IT solutions;
- The development of the Geoportal web system design
- Software was developed, and testing and debugging of the complex was conducted.

The implementation of these tasks will significantly improve monitoring and decision making on this issue.

## References

1. Arctic Human Development Report Regional Processes and Global Linkages: Nordic Council of Ministers, Copenhagen (2014). http://norden.diva-portal.org/smash/get/diva2:788965/FULLTEXT03.pdf. Accessed 20 Apr 2021
2. Barinov, M., Bolsunovskaya, M., Shirokova, S.: Software navigation system for the orientation of a transport robot on the ground. Transp. Res. Procedia **54**, 692–698 (2021)
3. Bolsunovskaya, M., Leksashov, A., Shirokova, S., Tsygan, V.: Development of an information system structure for photo-video recording of traffic violations. In: E3S Web of Conferences, p. 244 (2021)
4. Borovkov, A.I., Bolsunovskaya, M.V., Gintciak, A.M., Kudryavtseva, T.J.: Simulation modelling application for balancing epidemic and economic crisis in the region. Int. J. Technol. **11**(8), 1579–1588 (2020)
5. Chernogorskiy, S.A., Kozlov, A.V., Teslya, A.B.: Game-theoretic modeling of decision-making on state support for the infrastructure development in the Russian Far North. Int. J. Syst. Assur. Eng. Manag. **11**(1), 10–18 (2019). https://doi.org/10.1007/s13198-019-00798-6
6. Dmitrieva, T.G., Munkhalova, Y.A., Argunova, E.F., Alexeyeva, S.N., Egorova, V.B., Alexeeva, N.N.: The prevalence of chronic viral hepatitis in children and adolescents in Yakutia. Wiad. Lek. **68**(4), 553–556 (2015)
7. Economy of the North 2015 (ECONOR III): Sustainable Development Working Group Arctic Council Secretariat (2017). https://oaarchive.arctic-council.org/handle/11374/1946?show=full. Accessed 20 Apr 2020
8. Economy of the North (ECONOR) IV: Sustainable Development Working Group Arctic Council Secretariat. https://sdwg.org/what-we-do/projects/economy-of-the-north-econor/#:~:text=The%20ECONOR%20IV%20report%2C%20The,presented%20at%20Arctic%20Frontiers%202021.&text=The%20planned%20effect%20of%20ECONOR,Arctic%20sustainable%20development%20(Impact). Accessed 20 Apr 2021
9. Einarsson, N.: Arctic human development report (AHDR). In: Michalos, A.C. (ed.) Encyclopedia of Quality of Life and Well-Being Research. Springer, Dordrecht (2014). https://doi.org/10.1007/978-94-007-0753-5_104. Accessed 20 Apr 2021
10. Gerasimova, V.V., Maksimova, N.R., Levakoval, A., Zhebrun, A.B., Mukomolov, S.L.: Epidemiological features of chronic virus Hepatitis B and C in the Republic Sakha (Yakutia) Wiad. Lek. **68**(4), 502–507 (2015)
11. Gorbanev, S.A., et al.: Organization of an interregional monitoring system using GIS technologies by the example of Russian federation arctic zone. Gigiena i Sanitariya **97**(12), 1133–1140 (2018). https://doi.org/10.18821/0016-9900-2018-97-12-1133-1140

12. Gounder, P.P., et al.: Epidemiology of bacterial meningitis in the North American Arctic, 2000–2010. J. Infect. **71**(2), 179–187 (2015). https://doi.org/10.1016/j.jinf.2015.04.001
13. Heleniak, T., Turunen, E., Wang, S.: Demographic changes in the Arctic. In: Coates, K.S., Holroyd, C. (eds.) The Palgrave Handbook of Arctic Policy and Politics, pp. 41–59. Springer, Cham (2020). https://doi.org/10.1007/978-3-030-20557-7_4
14. Kandola, K., Chui, L., Li, V., Nix, N., Johnson, R., Case, C.: Examining DNA fingerprinting as an epidemiology tool in the tuberculosis program in the Northwest Territories, Canada. Int. J. Circum. Health **72**(1) (2013). https://doi.org/10.3402/ijch.v72i0.20067
15. Kozlov, A., Kankovskaya, A., Teslya, A.: Digital infrastructure as the factor of economic and industrial development: case of Arctic regions of Russian North-West. IOP Conf. Ser. Earth Environ. Sci. **539**(1), 012061 (2020)
16. Kovshov, A.A., Fedorov, V.N., Tikhonova, N.A., Novikova, Yu.A.: On the issue of digitalization in the field of ensuring the sanitary and epidemiological well-being of the population in the Russian Arctic. https://russian-arctic.info/info/articles/zdravookhranenie/opyt-sistem atizatsii-dannykh-o-sostoyanii-sanitarno-epidemiologicheskogo-blagopoluchiya-naseleniy a-a/. Accessed 18 Apr 2021
17. Mahnke, P.: LTS – Ubuntu Wiki (2017)
18. Odland, J.Ø., Donaldson, S.: Human exposure to pollutants and their health endpoints: the arctic perspective. In: Pacyna, J.M., Pacyna, E.G. (eds.) Environmental Determinants of Human Health. MIT, pp. 51–82. Springer, Cham (2016). https://doi.org/10.1007/978-3-319-43142-0_3
19. Omazic, A., Berggren, C., Thierfelder, T., Koch, A., Evengard, B.: Discrepancies in data reporting of zoonotic infectious diseases across the Nordic countries – a call for action in the era of climate change. Int. J. Circump. Health **78**(1) (2019). https://doi.org/10.1080/224 23982.2019.1601991
20. Pyankova, A.I., Fattahov, T.A.: Potential of the increase of life expectancy in the northern regions of Russia. Profilakt. Medit. **23**(2), 89–96 (2020). https://doi.org/10.17116/profmed20 202302189
21. Sleptsova, S.S., Semenova, V.K., Dyatchkovskaya, P.S., Nikitina, S.G.: Breadth spreading of viral Hepatitis markers in the risk groups in the Republic of Sakha (Yakutia). Wiad. Lek. **68**(4), 476–479 (2015)
22. Stella, E., Mari, L., Gabrieli, J., Barbante, C., Bertuzzo, E.: Permafrost dynamics and the risk of anthrax transmission: a modelling study. Sci. Rep. **10**(1) (2020). https://doi.org/10.1038/s41598-020-72440-6
23. van Stijn S.: Overview of Docker Compose. Docker Documentation (2021)
24. Uspenskij, M.B., Shirokova, S.V., Mamoutova, O.V., Zhvarikov, V.A.: Complex expert assessment as a part of fault management strategy for data storage systems. In: Arseniev, D., Overmeyer, L., Kälviäinen, H., Katalinić, B. (eds.) Cyber-Physical Systems and Control. CPS&C 2019. Lecture Notes in Networks and Systems, systems, vol. 95, pp. 592–600. Springer, Cham (2020). https://doi.org/10.1007/978-3-030-34983-7_58
25. Uspenskiy, M.B., Smirnov, S.V., Loginova, A.V., Shirokova, S.V.: Modelling of complex project management system in the field of information technologies. In: Proceedings of 2019 3rd International Conference on Control in Technical Systems, CTS 2019, pp. 11–14 (2019)
26. Yurev, V., Shirokova, S., Iliashenko, O., Dybok, D.: Analysis and evaluation of innovative development factors of the economy using technologies of the semantic web. In: Proceedings of the 33rd International Business Information Management Association Conference, IBIMA 2019: Education Excellence and Innovation Management through Vision 2020, pp. 9670–9676 (2019)
27. Zharova, M., Shirokova, S., Rostova, O.: Management of pilot IT projects in the preparation of energy resources. In: E3S Web of Conferences, p. 110 (2019)

28. API Reference for the ArcGIS API for Python — arcgis 1.8.5 documentation. https://develo pers.arcgis.com/python/api-reference/index.html. Accessed 20 Apr 2021
29. ArcGIS API for JavaScript. https://developers.arcgis.com/javascript/latest/. Accessed 23 Apr 2021
30. ArcGIS API for Python - ArcGIS for Developers [electronic resource]. https://developers.arc gis.com/python/. Accessed 23 Apr 2021
31. Europe entitled Barents Euro-Arctic Region (BEAR) Homepage. https://www.barentscoope ration.org/en. Accessed 20 Apr 2021
32. Empowering App Development for Developers, Docker. https://www.docker.com/. Accessed 20 Apr 2021
33. ESRI: What is the ArcGIS Enterprise portal? Portal for ArcGIS Documentation for ArcGIS Enterprise. https://enterprise.arcgis.com/. Accessed 20 Apr 2021
34. ESRI. About ArcGIS Web Adaptor—ArcGIS Enterprise Documentation for ArcGIS Enterprise. https://enterprise.arcgis.com/. Accessed 20 Apr 2021
35. ESRI. What is ArcGIS Data Store? — ArcGIS Enterprise Documentation for ArcGIS Enterprise. https://enterprise.arcgis.com/. Accessed 20 Apr 2021
36. GeoJSON – Справка ArcGIS Online – Documentation [electronic resource]. https://doc.arc gis.com/ru/arcgis-online/reference/geojson.htm. Accessed 21 Apr 2021
37. Minio documentation [electronic resource]. https://docs.min.io/docs/how-to-use-aws-sdk-for-php-with-minio-server.html. Accessed 18 Apr 2021
38. Nordic Council and the Nordic Council of Ministers Homepage. Copenhagen, Denmark. https://www.norden.org/en. Accessed 22 Apr 2021
39. Nordregio Homepage. https://nordregio.org/. Accessed 20 Apr 2021
40. Sustainable Development Working Group of the Arctic Council (SDWG). https://sdwg.org/about/. Accessed 18 Apr 2021, Accessed 22 Apr 2021
41. International Arctic Forum Homepage. https://forumarctica.ru/. Accessed 22 Apr 2021
42. The Russian Government Homepage. http://government.ru/en/. Accessed 22 Apr 2021

# AI-System for Predicting the Russia's GDP Volume Based on Dynamics of Factors in the Transport Sector

Nikolay Lomakin[1]([✉]) [iD], Anastasia Kulachinskaya[2] [iD], Uranchimeg Tudevdagva[3] [iD], Elena Radionova[4] [iD], and Natalya Mogharbel[1] [iD]

[1] Volgograd State Technical University, Volgograd, Russia
[2] Peter the Great St. Petersburg Polytechnic University, St. Petersburg, Russia
[3] Technische Universität Chemnitz, Chemnitz, Germany
[4] PRUE G.V. Plekhanov, Volgograd, Russia

**Abstract.** The article deals with the problem of artificial intelligence based forecasting the volume of Russia's gross domestic product (GDP). The fact that a decrease in the volume of exports and sales in the domestic market of the country in 2019 led to a decrease in the GDP and was accompanied by a drop in the goods turnover volume has been considered.

The scientific novelty lies in the fact that the study has put forward and confirmed the hypothesis that an artificial intelligence system makes it possible to obtain a predicted GDP value based on parameters of the real sector of economy, including the freight turnover indices, volume of transported goods, and factors characterizing the development of the financial sector.

Actual scientific works on artificial intelligence systems applied in the real sector of the economy in the context of digitalizing economy have been considered. Factors that determined the results of the industry in 2011–2019 have been analyzed.

The authors used a "perceptron" AI-system that consisted of an input layer, two hidden layers, and an output layer with the feature of the GDP volume and resulted in a predicted Russia's GDP value that amounted to 1610.3322 billion dollars at the end of 2020, which was less than its actual value by 0.05 billion dollars or 0.0027%.

**Keywords:** AI system · Cargo carriage volume · Perceptron · Transport industry · Forecast · GDP · Artificial intelligence

## 1 Introduction

The transport industry is an important sector of the Russian economy. This sector provides over 15% of Russia's GDP. The development of the industry is not stable. In 2019, there was a decrease in the volume of cargo transportation in the country against the background of a decrease in Russia's GDP due to many negative factors, including the COVID19 pandemic.

© Springer Nature Switzerland AG 2021
D. Rodionov et al. (Eds.): SPBPU IDE 2020, CCIS 1445, pp. 118–132, 2021.
https://doi.org/10.1007/978-3-030-84845-3_8

The originality of the study consists in the fact that the investigation formulated and proved the hypothesis that an artificial intelligence system enables obtaining a predicted GDP value based on the parameters of the real sector of economy, including indices of the freight turnover, the volume of transported goods and factors characterizing the development of the financial sector.

The relevance of the study is determined by the proved hypothesis that the developed artificial intelligence system enables obtaining a predicted value of the RF GDP volume. So was the novelty and practical relevance of our investigation.

The article presents a neural network model that allows getting a predicted RF GDP value based on the cargo carriage volume, freight turnover, and other factors.

The article discusses the problem of GDP forecasting to support managerial decision-making, ensuring the development of the transport sector in conditions of market uncertainty. In particular, issues of optimal interaction between transport modes, a need to increase investments in the transport industry, enhance their efficiency, and implement innovations of Industry 4.0 technologies remain urgent.

The reduction in exports and sales in the domestic market of the country in 2019 is known to lead to a decrease in the gross domestic product and resulted in a drop in the turnover volume.

The study used a "perseptron" AI-system that involved an input layer, two hidden layers, and an output layer with one feature—the RF GDP value that is the predicted value for the following year.

Forecasting the GDP volume of the Russian Federation is important, since it allows obtaining a more feasible forecast of the development in the real sector of the economy for the following year, as well as creating prerequisites for optimal interaction between sectors of the national economy.

Works of many Russian scientists are devoted to the issue of forecasting the turnover volume and GDP value in the Russian Federation; however, some aspects require further study. The article examines theoretical foundations of the artificial intelligence applied in the transport sector, considers the main ones that allow providing support for managerial decision-making in this sector.

The factors that determined the results of the industry in 2011–2019 were analyzed. We calculated the predicted GDP value in conditions of market uncertainty.

Statistical data were collected and processed; they formed the basis of the proposed analytical study that made it possible to identify current patterns and include making a neural network model of factorial features. Their dynamics largely determined changes in the effective feature—the GDP volume.

Thus, the results of calculations, involving processing large data and statistical indices of the country's transport industry for the period 2011–2019 formed the basis of the AI-system developed.

## 2 Literature Review

### 2.1 Artificial Intelligence Systems Applied B Transport Sector

The authors of the review of the Russian transport sector in 2019 consider the dynamics of the performance parameters and the factors that influenced the performance of certain modes of transport. The review of the world literature made it possible to name the experts, whose researches became a base for our work. Research studies showed that artificial intelligence systems are increasingly being used in all areas. The artificial intelligence implemented into the economy, finance, and transport is important in the context of digitalization of business processes. Specific features of various means of transport and logistics chains that ultimately determine the efficiency of the country's transport and its contribution to the GDP value should be noted.

The increase in road freight turnover that grew by 6% at the end of the year was explained by a drop in the cost of fuel and reduction in the transportation leg, which made it possible to redistribute part of the freight traffic from other modes of transport to road transport [1]. The development of the country's transport systems is increasingly considering advanced artificial intelligence, digitalization, and automation of business processes.

Many aspects of digitalization of the economy are considered in latest scientific works.

Liu, Gibson, and Osadchy in their work consider the application of deep machine learning on big data in the practical work of companies [2]. Udomsak applied computational models and compared the naive Bayesian classifier and the support vector machine with respect to their ability to predict the Thai stock exchange [3]. Shiralkar, Flam-mini, Menczer, and Ciampaglia noted necessity to find flows in knowledge graphs to support fact-checking [4].

The most important tendency in developing artificial learning systems applied in financial risk management is machine learning. For example, Breiman identified obstacles to machine learning [5]. Baltas et al. investigated the problem of stock selection, using machine learning [6].

Gromova's study of digitalization applied in all areas, including the digital economy with an emphasis on the automotive industry in Russia [7], procedure for modifying Kulagina's methods for assessing the innovation system [8, pp. 5083–5091], and Gutman's development of the northern sea route in the system of international transport corridors [9] are of great practical interest.

Experience has proven that increasing the accuracy of predicting the GDP volume based on the artificial intelligence makes it possible to create prerequisites for ensuring the sustainability of economic development, both on a global scale and at the levels of regions and enterprises. Babskova et al. examined the innovation activities, developing in the region in the context of the digital environment [10, pp. 4361–4365]. Zaborovskaya; Kudryavtseva; Zhogova et al. studied mechanisms of a regional strategy for sustainable development on the example of the Leningrad region [11, pp. 5065–5076].

The artificial intelligence used for solving a number of practical problems in the field of transport play an important role in modern conditions. Vychytilova et al. investigated macroeconomic factors that explain the stock price volatility based on empirical data of the automotive industry in many countries [12, pp. 3327–3341].

Algorithms based on data mining were studied by Kumar, Kumar, Durg, and Chauhan and are also of great importance [13].

## 2.2 Theoretical Foundations for Modeling the Transport Sector

Macroeconomic factors included in the Vychytilova's AI model for explaining the stock volatility in the automotive industry require assessment and accounting of financial risk [14]. The authors compared various mathematical models.

Relevant is the study of protection against financial risks in the stock market. Although many studies were devoted to the issues of financial risk management, certain aspects require further scientific study.

Ruppert's research study of statistics and data analysis for financial engineering [15] and Jensen et al. capital asset pricing model [16] are also of scientific interest.

Fama and MacBeth found it necessary to consider risk as a category that experiences return and equilibrium [17], while Zhang Lu suggested focusing on the cost premium, being also important [18]. Frazzini and Pedersen took into account the role of beta portfolio of financial instruments [19].

# 3 Materials and Methods

Monographic method, calculation method, and a "perseptron" neural network were used in the work. The neural network was created on the Deluctor platform.

Actual scientific works on artificial intelligence systems applied in the real sector of the economy in the context of digitalizing economy have been considered. Factors that determined the results of the industry in 2011–2019 have been analyzed.

The results of calculations, involving processing large data and statistical indices of the country's transport industry for the period 2011–2019 formed the basis of the AI-system developed.

The neural network algorithms and macroeconomic parameters applied in modeling the transport sector also matter. Factor models enable analyzing the sensitivity of asset returns, depending on one or more factors, as established in the studies conducted by Dubravka and Posedel [20]. They compared three models of mixed effects as follows. All models contained three levels. The car manufacturer was considered the first tier, the country was the second tier, and time was the third one. The time was used as a random factor; country and car manufacturer were random variables. The authors also applied the AR (1) correlation structure to model first-order autoregression,

$$Y_{i,j,k} = \beta_{0,j,k} + \beta_{1,j,k}x_1 + \beta_{2,j,k}x_2 + \ldots + \beta_{n,j,k}x_1 + \varepsilon_{i,j,k}$$

where:

$$\varepsilon_{i,j,k} = N(0, \Sigma) \tag{1}$$

In order to calculate the predicted value, the authors used a perceptron; its artificial neuron is presented below (Fig. 1).

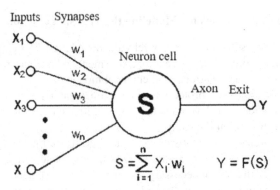

**Fig. 1.** Artificial neuron

Each neuron is characterized by its current state by analogy with nerve cells in the brain that can be excited or inhibited. It has a group of synapses—unidirectional input connections to the outputs of other neurons, and also has an axon—the output connection of this neuron.

Each synapse is characterized by the size of the synaptic connection or its weight $wi$. The current state of the neuron is defined as the weighted sum of its inputs:

$$s = \sum_{i=1}^{n} x_i \cdot w_i \tag{2}$$

The output of the neuron is a function of its state:

$$y = f(s) \tag{3}$$

The nonlinear function $f$ is called "activation function" and can have various forms. One of the most common is a non-linear function with saturation, the so-called logistic function or sigmoid (i.e., an S-shaped function):

$$f(x) = \frac{1}{1 + e^{-\alpha x}} \tag{4}$$

It is obvious in the sigmoid expression that the output value of the neuron lies in the range of [0.1].

When processing not very large amounts of data, it is advisable to use an elementary perceptron, in which all elements are simple. In this case, its activating function has the form:

$$c_{ij}(t) = U_i(t-\tau)\, w_{ij}(t) \tag{5}$$

In order to generate a neural network, there was formed a data set that included necessary natural and cost indices for the period under study (Table 1).

**Table 1.** Dataset for a neural network

| Freight turnover million tons / km | Aeroflot stock | UTAR | FESH | NMTP | FESCO | RTS Index | USD | Railway, billion tons | Auto, billion tons | Pipe, billion tons | Brent | GDP |
|---|---|---|---|---|---|---|---|---|---|---|---|---|
| 4245 | 106,78 | 7,17 | 8,86 | 9,05 | 7,72 | 1517 | 61,98 | 1399 | 5735 | 608 | 63,06 | 1610,38 |
| 4189 | 112,06 | 7,7 | 5,01 | 6,89 | 4,63 | 1068 | 69,83 | 1411 | 5544 | 603 | 56,98 | 1630,66 |
| 4063 | 142,12 | 8,9 | 6,1 | 8,12 | 5,97 | 1154 | 57.61 | 1384 | 5404 | 589 | 64,05 | 1578,41 |
| 3900 | 160,06 | 8,9 | 3,26 | 6,53 | 3,11 | 1152 | 61,27 | 1325 | 5397 | 578 | 54,44 | 1282,66 |
| 3821 | 57.2 | 11,26 | 2,64 | 3,51 | 2,71 | 757 | 73,6 | 1329 | 5357 | 578 | 37.72 | 1363,7 |
| 3768 | 34,52 | 8,56 | 2,99 | 1,6 | 2,85 | 790 | 55,91 | 1375 | 5417 | 566 | 62,16 | 2056,58 |
| 3670 | 73,56 | 23,23 | 3,91 | 2,78 | 3,93 | 1442 | 32,89 | 1381 | 5635 | 558 | 110,63 | 2289,24 |
| 3471 | 45,08 | 24,53 | 9,65 | 2,9 | 9,44 | 1526 | 30,56 | 1421 | 5842 | 555 | 109,64 | 2202,67 |
| 3333 | 51,4 | 24,53 | 8,34 | 2,8 | 8,51 | 1381 | 32.2 | 1382 | 5663 | 576 | 107,97 | 2044,61 |

It seems expedient to use an elementary perceptron to form predicted values for the volume of Russia's GDP, depending on the dynamics of freight turnover.

## 4   Results and Discussion

### 4.1   The Russia's GDP Volume Depending on the Freight Turnover Dynamics

The scientific originality is that our research study put forward and proved the hypothesis that the artificial intelligence system allows obtaining a predicted GDP value that considers the parameters of the real sector of the economy, including the freight turnover indices, volume of transported goods, and factors characterizing the development of the financial sector.

The research evidenced that there is a certain relationship between the RF GDP value and the turnover volume (Fig. 2).

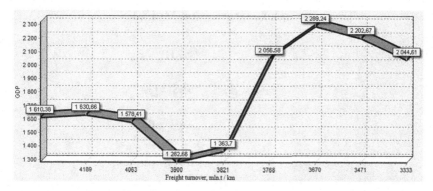

**Fig. 2.** Dynamics of the GDP volume of freight turnover in the Russian Federation

There is a downward trend in GDP as the trade turnover grows. For example, for the period of 2011–2019, the freight turnover increased from 3333 million tons to 4245 million tons or by 127.36%, while the GDP dropped from 2044.61 billion dollars to 1610.38 billion dollars or down to 78.76%.

The structure of transported goods did not change significantly during the period under review (Fig. 3).

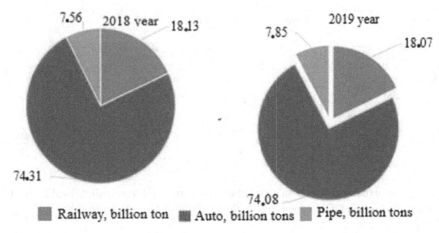

**Fig. 3.** The dynamics of transported goods structure, 2018–2019

In 2019, the largest share in the structure of transported goods is accounted for by road transport of 74.08%, rail transport of 18.07%, and pipeline transport of 7.85%.

The sustainable development of the country's economy is largely determined by the rhythmic work of transport, with the planning of the GDP volume depending on many factors. A nonlinear mathematical neural network model seemed expedient to form.

The dynamics of the share price of transport companies—the leaders in the growth and fall of the Transport Index (MOEXTN) for February 28, 2021 is presented in the chart below (Fig. 4).

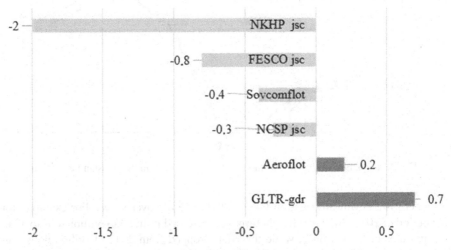

**Fig. 4.** Dynamics of the share price of transport companies

It seems expedient to develop an AI system Perceptron based on the initial data.

## 4.2 Perceptron AI System

The model used data representing the performance indices of the country's transport industry in dynamics for the period of 2011–2019, i.e. the volume of the country's cargo turnover; goods transported by rail, road, and pipeline transport; shares of transport companies Aeroflot, UTAR, FESH, Novorossiysk seaport (NMTP), and Far Eastern Shipping Company (FESCO); and parameters, reflecting the dynamics of the financial sector, including the RTS index, dollar exchange to the Russian ruble (USD), and the price of oil (brent). The proposed AI-system enabled obtaining a predicted value of the Russia's GDP volume for 2020.

The work resulted in a perceptron; its graph is presented below (Fig. 5).

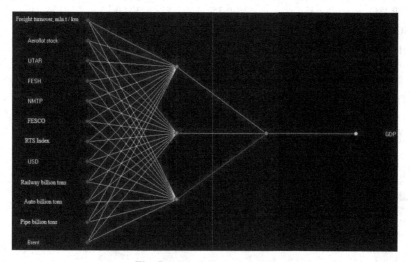

**Fig. 5.** Neural network graph

The neural network graph reflected the structure of the proposed artificial intelligence system that contained an input layer with 12 parameters, two hidden layers, and an output layer with one parameter.

## 4.3 GDP Forecast Based on "What-If" Function

Our research resulted in the "perseptron" AI system that allowed calculating the forecast of the GDP growth.

The model dataset included data reflecting the performance of the country's transport industry for the period of 2011–2019. The input layer of the model involved the volume of the country's cargo turnover; volume of goods transported by rail, road, and pipeline transport; shares of the transport companies Aeroflot, UTAR, FESH, Novorossiysk Sea Port (NMTP), and Far Eastern Shipping Company (FESCO); and parameters, reflecting the dynamics of the financial sector, including the RTS index, the dollar exchange to the Russian ruble (USD), and the price of oil (brent). The output layer had the only parameter—the forecast GDP value, billion dollars.

Moreover, the model included stock quotes of some key companies in the transport industry (Table 2).

**Table 2.** Stock quotes of leading transport companies

| Bank | 28.02.2021 Share price, RUB | Change per day, % | Volume, RUB mln. | Capitalization, RUB bln. |
|------|------|------|------|------|
| AFLT | 69.58 | 0.20 | 769.91 | 2.8 |
| GLTR:TQBR | 497.65 | 0.66 | 74.33 | 0 |
| FLOT | 90.2 | −0.35 | 85.17 | 214.22 |
| NMTP JSC | 8.03 | −0.25 | 30.66 | 154.66 |
| FESCO JSC | 12.06 | −0.82 | 22.92 | 35.59 |
| NKHP JSC | 359.5 | −0.04 | 1.99 | 24.30 |

The quality of the generated mathematical AI-model was considered high, since none of the values of the parameters included in the model went beyond the confidence intervals (Fig. 6).

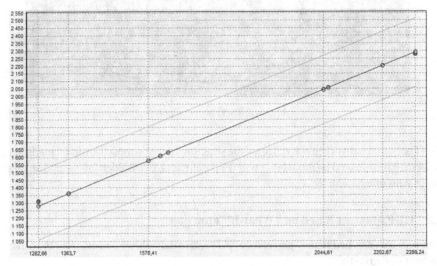

**Fig. 6.** Scatter plot

The neural network was of high quality and made it possible to obtain accurate predictions with minimum errors.

Based on the "what-if" function in Deduktor, the predicted values of growth in bank assets were obtained in terms of the banks under study (Fig. 7).

| Field | Value |
|---|---|
| ⊟ 🔓 Input | |
| 9.0 Freight turnover, mln.t / km | 4245 |
| 9.0 Aeroflot stock | 106.78 |
| 9.0 UTAR | 7.17 |
| 9.0 FESH | 8.86 |
| 9.0 NMTP | 9.05 |
| 9.0 FESCO | 7.72 |
| 9.0 RTS Index | 1517 |
| 9.0 USD | 61.98 |
| 9.0 Railway, billion tons | 1399 |
| 9.0 Auto, billion tons | 5735 |
| 9.0 Pipe, billion tons | 608 |
| 9.0 Brent | 63.06 |
| ⊟ 🔓 Output | |
| 9.0 GDP | 1610.33227161402 |

**Fig. 7.** The GDP forecast based on the "what-if" function

If we substitute the values of the last time interval into the neural network model, the forecast RF GDP value is obtained. At the end of 2020, it was 1610.3322 billion dollars, which was by 0.05 billion dollars or 0.0027% less than its actual value.

## 5  Discussion

### 5.1  Current State of the Transport Sector

We should recognize that the proposed approach does not solve the entire set of problems concerning the forecast value of the RF GDP; its application does not cause the replacement of existing methods based on the balances of raw materials, physical resources, commodity, labor resources, and others. However, in the conditions of market uncertainty and turbulent economy, the proposed approach has scientific value, since it provides an opportunity to use the advantages of artificial intelligence systems. The predictive functions of AI-systems ensure the achievement of a certain economic result—the predicted GDP value at different levels the real sector of the economy, in particular, its transport sector.

Our investigation is based on approaches of such scientists as De Meo and Tizzanini [21]. The authors proposed a tool for predicting risks to economic growth and spillovers from international business cycles by considering gross domestic product (GDP) as a conditional value at risk (CoVaR).

Through his scientific research, Brian Tarran came to the conclusion that the GDP shows the size of a country's economy, but says little about the economic well-being of citizens. The author agreed with experts who discussed new ways of measuring what matters [22].

Simes & O'Mahony assessed the long-term economic impact of digital technologies on the GDP and considered the level of the mobile phone penetration and Internet use as general indicators. Their empirical results showed that the spread of digital technologies has significantly improved production volumes in Australia and other countries, contributing to a steady increase in gross domestic product per capita by an average of about 5.8% [23].

Among many factors affecting the GDP, cyclical fluctuations, financial shocks, and emergence of fast-growing entrepreneurial start-ups should be emphasized. Christoph & Caggese found that cumulative negative financial shocks reduce all types of startups, but their impact is much stronger for startups with high growth potential, especially at low GDP growth. The obtained results revealed a new composition of the input channel that considerably reduces the employment growth and is potentially important in explaining the slow recovery from financial crises [24].

In the context of a slowdown in economic growth due to the lockdown of almost all industries during the COVID-19 pandemic, OPEC + crisis, fall in oil prices, and unfavorable conditions on world markets, the stagnation of Russian transport cargo turnover occurred for the first time since 2015. The water and air transport segments suffered the most from the slowdown despite the growth in foreign trade. At the end of 2019, the rail transportation showed stagnation, as the growth in export traffic was offset by a drop in domestic volumes.

Such challenges force market participants to reconsider approaches, find new points of growth for their business, and rebuild current supply chains, relying on unmanned modes of transport and other solutions in the context of a new technological paradigm and Industry 4.0.

However, as evidenced in practice, many of the current problems in this sector remain unresolved. For example, a high dead mileage reduced the efficiency of cargo transportation and led to congestions on key routes. The problem of timely replacement of obsolete vehicles in many modes of transport, except for railway and air transport, has not been resolved; the digitalization is being performed at an inappropriate pace; the artificial intelligence is limited by the sphere of internal corporate analytical systems. The infrastructure is far from perfect in terms of handling cargo traffic. Macroeconomic factors have a powerful impact on the transport industry.

A decrease in growth rates in 2019 was observed in many basic industries. Manufacturing that accounted for the most significant contribution to GDP growth had a lower value than a year earlier by 1.2 sub-points (2.3% versus 3.5%); retail trade slowed to 1.9% after 2.8% in 2018; construction grew by only 0.6% [25].

The slowdown in growth was noted throughout the world and led to a decrease in the growth rate of the world GDP (from 2.9 to 2.2%) and a slowdown in the growth rate of the world trade (from 3.7 to 0.8%). Due to the world market movement, there was a drop in the physical quantity of supplies and a decrease in the export indicator by 6% compared to the previous year.

The studies showed that negative trends were especially noted in the export of ferrous metals and wheat, since there was a decrease in the factor value that made 12.6% and 27.6% by volume and 22.4% and 24.1% in monetary terms [26]. Moreover, insignificant positive dynamics was observed only in the export of coal, crude oil, and fertilizers, but only fertilizers grew in monetary terms.

A sharp decline in Russian exports is known to be determined not only by stagnation in the world trade, but also by other factors that led to changes in the balance of payments and an increase in capital outflow.

In 2019 there was a drop in the current account surplus of the country's balance of payments and an increase in the net capital outflow from Russia. However, in general, the situation still remained at a relatively normal level against the background of historical indices. According to the Central Bank of the Russian Federation, the volume of net lending by the private sector of the Russian Federation decreased by 65.6% down to $ 21.9 billion, which was caused by a decrease in foreign liabilities of banks and a drop in the size of deposits in lending to the rest of the world by other sectors [27].

At the end of 2019, the population's monetary income increased by 1%, which was caused by a slowdown in consumer price growth, with the real disposable income of the population returning to growth despite a rapid decline over three years up to 2018 [28].

It is consumption that remains the main factor in GDP growth, although consumption in 2019 was assessed as weak. At the same time, the issue of transition to an investment model that involves not only accelerating investments, but also increasing their efficiency due to the automation of business and artificial intelligence will be dealt with.

The country's investment quote to GDP was 25% in the beginning of 2018 and then grew to 27%, i.e. it did not practically change for almost a decade despite the directives set in May decrees 2012 and 2018 to increase it [29].

According to the latest data from the Federal State Statistics Service of the Russian Federation, the share of investments in fixed assets in GDP amounted to 20.6% in 2019, which was by 0.6 points higher than updated indices for 2018 [30].

The investments of enterprises in fixed assets were noted to be increased by 1.7% to 19.3 trillion rubles, slowing down in comparison with 2018 [31].

In modern conditions, it is important to conduct research in the field of modeling and implementation of innovations into the real sectors of the economy. The research study conducted by Rudskaya and Rodionov in the field of econometric modeling as a tool for assessing effectiveness of regional innovation systems [32], Nikolova's work [33] that considered the impact of globalization on the risk assessment of innovation projects, and Kolesnikov's study [34] of bank loans as a source of financing for innovative enterprises also draw attention.

## 6   Conclusions

A. The artificial intelligence used to support managerial decision-making in conditions of market uncertainty is of great importance. The latest scientific works regarding the artificial intelligence systems applied for the development of digital economy were investigated. In particular, the issue of optimal interaction between modes of transport, the need to increase investments in the transport industry and enhance their efficiency, as well as the implementation of innovations in line with Industry 4.0 technologies in transport remain important.

B. In the research study conducted, factors that determined the performance of the transport industry in 2011–2019 were analyzed. There was put forward and proved a hypothesis that an artificial intelligence system makes it possible to obtain the predicted value of Russia's GDP due to parameters in the neural network model that reflect the development of the real sector of the economy and include the dynamics of goods turnover, volume of transported goods, and some indices characterizing the financial sector, i.e. the dollar exchange rate and the price of brent oil.

C. A "perceptron" AI-system that consisted of an input layer, two hidden layers, and an output layer with one feature—the GDP volume—was generated. The quality of the obtained AI-model was considered high, since none of the values of the parameters in the model went beyond the confidence intervals. As a result of substitution of the input parameters values of the last time interval into the neural network model, the forecast RF GDP value was obtained. At the end of 2020, it amounted to 1610.3322 billion dollars, which was less than its actual value by 0.05 billion dollars or 0.0027%.

# References

1. Overview of the Russian transportation sector in 2019. KPMG in Russia and the CIS (2020). https://ru.investinrussia.com/data/file/ru-ru-russian-transport-survey.pdf. Accessed 24 May 2021
2. Liu, S., Gibson, J., Osadchy, M.: Learning to Support: Exploiting Structure Information in Support Sets for One-Shot Learning. LG, cs.AI, stat.ML (2018). https://deepai.org/public ation/learning-to-support-exploiting-structure-information-in-support-sets-for-one-shot-lea rning. Accessed 24 May 2021
3. Udomsak, N.: How do the naive Bayes classifier and the Support Vector Machine compare in their ability to forecast the Stock Exchange of Thailand? cs.LG (2015). https://arxiv.org/ftp/arxiv/papers/1511/1511.08987.pdf Accessed 24 May 2021
4. Shiralkar, P., Flammini, A., Menczer, F., Ciampaglia, G.L.: Finding Streams in Knowledge Graphs to Support Fact Checking. cs.AI, cs.SI (2017). https://ieeexplore.ieee.org/document/8215568. Accessed 24 May 2021
5. Breiman, L.: Bagging predictors. Mach. Learn. pp. 123–140 (1996).
6. Baltas, N., Jessop, D., Jones, C., Lancetti, S., Winter, P., Holcroft, J.: Low-risk investing: perhaps not everywhere. Quantitative Monographs. UBS Global Research (2015). https://www.researchgate.net/publication/338964209. Accessed 24 May 2021
7. Gromova, E.A.: Digital economy development with an emphasis on automotive industry in Russia. Espacios **40**(6) (2019). https://www.scopus.com/record/display.url?eid=2-s2.0-850 61654343&origin=resultslist. Accessed 24 May 2021
8. Kulagina, N.A., Mikheenko, O.V., Rodionov, D.G.: Technologies for the development of methods for evaluating an innovative system. Int. J. Recent Technol. Eng. **8**(3), 5083–5091 (2019)
9. Gutman, S., Konnikov, E., Kuznetsov, R.: The Northern Sea Route: problems and development potential in international transport corridors. Espacios **40**(25) (2019). https://www.scopus.com/record/display.url?eid=2-s2.0-85071174293&origin=resultslist. Accessed 24 May 2021
10. Babskova, O., Nadezhina, O., Zaborovskaya, O.: Innovative activities in a region in the conditions of the development of the digital environment. Int. J. Innov. Technol. Exp. Eng. **8**(12), 4361–4365 (2019)
11. Zaborovskaya, O., Kudryavtseva, T., Zhogova, E.: Examination of mechanisms of regional sustainable development strategy as exemplified by the Leningrad region. Int. J. Eng. Adv. Technol. **9**(1). 5065–5076 (2019)
12. Vychytilová, J., Pavelková, D., Pham, H., Urbánek, T.: Macroeconomic factors explaining stock volatility: multi-country empirical evidence from the auto industry. Econ. Res. Ekon. Istraž. **32**(1), 3327–3341 (2019)

13. Kumar, P., Kumar, N.V., Durg, S., Chauhan, S.: A Benchmark to Select Data Mining Based Classification Algorithms For Business Intelligence And Decision Support Systems. .DB, cs.LG (2012). https://pdfslide.net/documents/a-benchmark-to-select-data-mining-based-cla ssification-algorithms.html. Accessed 24 May 2021

14. Vychytilovaa, J., Pavelkovaa, D., Phamb, H., Urbaneka, T.: Macroeconomic factors explaining stock volatility: multi-country empirical evidence from the auto industry. Econ. Res. ekonomska istrazivanja **32**(1), 3327–3341 (2019)

15. Ruppert, D.: Statistics and Data Analysis for Financial Engineering. Springer (2019). https:// www.springer.com/gp/book/9781493926138. Accessed 24 May 2021

16. Jensen, M., Fischer, V., Myron, V.: The Capital Asset Pricing Model: some empirical tests. Praeger Publishers Inc. (1972). https://www.maths.usyd.edu.au/u/UG/IM/MATH2070/r/Bla ckJensenScholes_StudiesInTheTheoryOfCapitalMarkets1972.pdf. Accessed 24 May 2021

17. Fama, E.F., James, D.M.: Risk, return and equilibrium: empirical tests. J. Polit. Econ. **81**(3). 607–636 (1973)

18. Lu, Z.: The value premium. J. Fin. 67–103 (2005). https://studylib.net/doc/10987670/ the-value-premium-journal-of-finance---2005--60--1---67%E2%80%931. Accessed 24 May 2021

19. Frazzini, A., Pedersen, L.H.: Betting against beta. NBER Working Paper . https://www.nber. org/papers/w16601. Accessed 24 May 2021

20. Dubravka, B., Posedel, P.: Do macroeconomic factors matter for stock returns? Evidence from estimating a multifactor model on the Croatian market. Bus. Syst. Res. **1**(1–2), 39–46 (2010). https://doi.org/10.2478/v10305-012-0023-z

21. De Meo, E., Tizzanini, G.: GDP-network CoVaR: a tool for assessing growth-at-risk J. Fin. **23**(1) (2021). https://doi.org/10.1111/ecno.12181

22. Tarran, B.: Is it time for GDP to be dethroned? J. Fin. (2018). https://doi.org/10.1111/j.1740-9713.2018.01121.x. Accessed 24 May 2021

23. Simes, J., O'Mahony, J.: How do digital technologies drive economic growth? J. Fin. (2017). https://doi.org/10.1111/1475-4932.12340

24. Christoph, A., Caggese, A.: Cyclical fluctuations, financial shocks, and the entry of fast-growing entrepreneurial startups. Rev. Fin. Stud. (2020). https://doi.org/10.1093/rfs/hhaa112. Accessed 24 May 2021

25. Business activity pattern for Q1 2020: Ministry of Economic Development, April 30. https://openknowledge.worldbank.org/bitstream/handle/10986/34219/Russia-Recession-and-Growth-Under-the-Shadow-of-a-Pandemic.pdf?sequence=4. Accessed 24 May 2021

26. Customs statistics of Russia for January-December 2019. Federal Customs Service of the Russian Federation (2020). https://customs.gov.ru/press/federal/document/267169. Accessed 24 May 2021

27. Assessment of the balance of payments in the Russian Federation for Q II 2020, the Central Bank of the Russian Federation. https://cbr.ru/Collection/Collection/File/29131/Bal ance_of_Payments_2020-02_4_e.pdf. Accessed 24 May 2021

28. Statistics on monetary income of the population in 2019 (2020). FSSS RF http://www.kre mlin.ru/acts/bank/42902. Accessed 24 May 2021

29. Presidential Address to the Federal Assembly, March 1, 2018. http://www.consultant.ru/doc ument/cons_doc_LAW_291976/. Accessed 24 May 2021

30. Equity contribution in fixed assets in GDP and UISIS. https://rosstat.gov.ru/folder/11186?pri nt=1. Accessed 24 May 2021

31. Socio-economic situation in Russia, Report, FSSS RF, January 2020. https://rosstat.gov.ru/ compendium/document/50801. Accessed 24 May 2021

32. Rudskaya, I., Rodionov, D.: Econometric modeling as a tool for evaluating the performance of regional innovation systems (with regions of the Russian Federation as the example). Academy of Strategic Management Journal (2017). https://www.semanticscholar.org/author/ I.-Rudskaya/122531390. Accessed 24 May 2021

33. Nikolova, L.V., Rodionov, D.G., Afanasyeva, N.V.: Impact of globalization on innovation project risks estimation. Eur. Res. Stud. J. **XX**(2B), 396–410 (2017)

34. Kolesnikov, A.M., Dubolazova, Y.A., Yakishin, Y.V.: Bank loans as a source of financing innovation enterprises. In: Proceedings of the 32nd International Business Information Mangement Association Conference, 15–16 November 2018, Seville, Spain. pp. 5496–5499 (2018)

# A System of Autonomous Control of Robotic Devices Based on an Embedded Virtual Model on the Example of a Symmetrical Walking Robotic Platform

Vasilyanov Georgiy$^{(\boxtimes)}$ ⓘ and Vassiliev Alexei ⓘ

Institute of Computer Science and Technology, Peter the Great St. Petersburg Polytechnic University, 195251 Saint Petersburg, Russia

**Abstract.** Modern robotics is a dynamically developing direction in science and industry Robots themselves are able to solve a very wide range of tasks depending on the field of application.

One of the most popular applications of robotic solutions is the search and rescue industry. Here, more than ever, the equipment able to move autonomously in conditions that are dangerous or deadly for humans are in demand.

The most promising direction of development of robotics in this area is the creation of autonomous walking robots, which, unlike wheeled robots, are able to move under conditions of a priori uncertainty of the environment. Moving in such an environment requires a complex, intelligent control system.

The goal of this work is to improve the efficiency of functioning of mobile autonomous robotic systems through application of new algorithmic and design solutions in their control systems.

The authors propose changes to the structure of the embedded intellectual control system, expressed in the introduction of a predictive control module at the strategic level, interacting with a virtual model of the robotic complex. The robotics system will become capable of solving predictive control problems when, in parallel with the real-time control system, the operation of simulation software is possible, which will allow, based on the information about the environment and the current state of the robotic complex, to "predict" the future behaviour of the robot.

In addition, the authors propose the concept and design of a symmetrical walking robotic platform based on a symmetrical three-dimensional propulsion system. The main difference from existing solutions lies in the special patented design of the propulsion system, which allows the robotic platform, made on its basis, to function both in the normal and in the overturned state without loss of functionality. A prototype was developed.

Using these solutions will allow:

1. Exploring the potential of adaptive control algorithms.
2. Reducing the development time of robotic complexes applying the portable virtual model.
3. Significantly increasing the ability of robotic complexes to overcome irregularities and heterogeneity of the surfaces due to the new design solutions.

© Springer Nature Switzerland AG 2021
D. Rodionov et al. (Eds.): SPBPU IDE 2020, CCIS 1445, pp. 133–148, 2021.
https://doi.org/10.1007/978-3-030-84845-3_9

**Keywords:** Control system · Walking robot · Model · Controller · Predictive algorithms

# 1 Introduction

Robotics is one of the most dynamically growing areas of technology, driven by the ever-increasing need for partial or complete automation of a broad class of management processes.

Many robotic systems at the current stage of development are divided into the following basic classes:

1. Industrial robots designed to perform routine operations, such as in factories or warehouses.
2. Military robots designed to perform various kinds of combat tasks, such as information gathering, demining, evacuation from the battlefield.
3. Search and rescue robots designed to search for and evacuate people in technogenic or natural accidents, to perform operations eliminating the consequences of this kind of accidents.
4. Household robots designed to help people in housekeeping.
5. Entertainment robots designed for interaction with humans during various events.

Of course, this is not an exhaustive list of classes in which robots have found use. They can be found in almost any sphere of modern life.

It is worth singling out one important circumstance: robots are able to perform their tasks in conditions that are often incompatible with the conditions of human life. When supplied with energy and timely service, they are able to work in hazardous industries, in dangerous environments, while working longer and often better than humans. In this regard, one of the most popular applications of robots was the field of search and rescue.

Since 2001, when the PackBot robots were first used in the aftermath of the 9/11 attacks [10], search and rescue robots have been a priority development direction in the world.

Over the past nearly twenty years, search and rescue robots have been used to deal with the effects of a large number of accidents, but one of the most significant milestones in their history has been the use of robots to deal with the aftermath of the Fukushima nuclear power plant accident [15]. Many different robots were used to deal with debris, for information collection and handling radioactive materials. A case that took place during one of the operations is indicative: for example, several robots were lost due to the fact that a cable was broken or the radio signal of communication with the operator disappeared. As a result, the robots were irretrievably lost.

Today, interesting in autonomous robots for search and rescue operations are large enough, which is confirmed by various research works on this topic [1, 3, 7, 14]. Developments are carried out both in the field of finding new design solutions and in the field of control systems.

As we can see, modern conditions impose increased requirements to the autonomy and intelligence of control systems. Intellectualization and the need to provide the robot with the ability to independently plan and realize movement is an urgent task of modern robotics.

Constant competition and the growing speed of development of the market for robotic systems require developers to search for both means of intellectualization of the developed robots and means of accelerating the development process. At the moment, with the growth of the computing power of embedded control systems, it becomes possible to transfer complex calculations from a remote computational control terminal to the robot itself.

At the same time, a significant expansion of modeling capabilities and an increase in computing power leads to the fact that modeling of robotic solutions is increasingly used to develop new robots [5, 8, 9].

The fastest growing and most popular class of modern robots are walking robots. They have higher manoeuvrability on an unprepared surface in comparison with other types of robotic devices. Adaptive control systems of such robots allow to improve the quality of robotic systems: solving problems of adaptation to irregularities and heterogeneity of the terrain, tasks related to overcoming or bypassing obstacles, tasks of choosing manner of moving and its trajectory, etc.

## 2 Literature Review

Today, the most famous robots are developed by Boston Dynamics and the Massachusetts Institute of Technology (MIT). Among them, we can single out such robots as the BigDog, Cheetah, LS3, Spot and others. At the same time, many companies around the world are trying to develop similar solutions based on these design [12], such as the Laikago robot developed by Unitree Robotics in China or the Tsuki robot created by Lingkang Zhang, or Little Elephant Robot [6]. We should note that both Boston Dynamics and MIT adhere to the principle of dividing areas of responsibility between a robot and a human, according to which a human is responsible for creating a functioning goal, while a robot must take on the task of achieving the goal set by a human [4].

It should be noted that interest in various kinds of predictive algorithms is relevant and natural, and this is due to the fact that when driving in conditions, associated with a dynamic change of the situation, it is important to predict where a potentially dangerous object may move. Most often, this is the prediction of the trajectories of the robot's movement to avoid an obstacle [16] or control various actuators [11]. In this case, different models are used for different tasks, while an approach can be used in which the model for the simulation remains one.

It is obvious that the task of achieving the goal is most often connected with the movement along the route or to the specified point; performing it is impossible without the use of adaptive algorithms and intelligent control systems.

Taking into account the fact that search and rescue robots operate in conditions of uncertainty, where dynamic changes of the environment are possible, we can highlight a number of problems associated with the formation of modern robotic solutions for this area:

1. When controlled by remote control (operator control) there is a risk of loss of communication and, as a result, loss of the robot. Accordingly, there arises a need to create autonomous robots;
2. The control systems of autonomous robots are quite complex and the requirements for such systems are very high. Therefore, there is a need to improve control systems, as well as to introduce predictive control capabilities;
3. Due to the widespread use of robotics, there is constant competition among the developers of robotics complexes. Consequently, it is necessary to reduce the time it takes to develop the finished product.

The goal of this work is to improve the efficiency of mobile autonomous robotic systems through the application of new algorithmic and structural solutions in their control systems.

To achieve this goal, the following tasks must be solved:

1. Development of the structure of the trajectory planning system;
2. Development of decision-making schemes in the face of a priori uncertainty of the environmental conditions;
3. Development of the design of a prototype of a walking robotic device for carrying out field experiments.

## 3    Methods

Since moving on an unprepared surface requires adaptability, the control system for such a robot has three main levels of control: strategic, tactical, and executive. In order for a control object to complete a task, it needs a functioning goal set by a control subject (such as an operator). At the same time, the goal can change or adjust dynamically.

- The goal of the strategic level is to determine the strategy of achieving the goal. For example, if we talk about the movement of the robotic platform (RP), the strategic level determines the set of stages (tactics) of movement along the route.
- Based on the information from the strategic level, the second, tactical, level determines the totality of elements of movement at the next stage.
- The task of the lowest, executive, level is to work out the totality of the elements of movement by generating signals for executive devices.

Each of the levels of control has connections with neighbouring levels. These levels together are an intelligent control system. At the same time, the control system implemented on an embedded control device is called an embedded intelligent control system or EICS.

The EICS is a technical object (process) management system integrated with it, implemented on the basis of embedded computing tools and using intelligent decision-making techniques [2].

Figure 1 presents the general structure of the EICS is shown on Fig. 1:

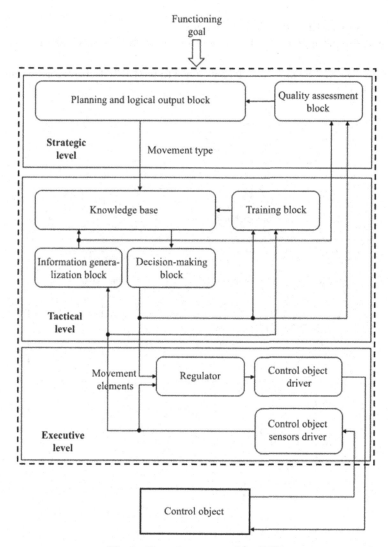

**Fig. 1.** General structure of the EICS

We should note that the existing model has a number of drawbacks, in particular:

1. Lack of means of predictive modelling of the external environment;
2. Lack of means of predicting the robotic object's own actions.

In order to improve the quality of intelligent control of the robotic complex, we propose the concept consisting of:

1. Changes to the EICS decision-making model;
2. Changes to the approach to designing a robotic complex control system.

Further, we will consider the changes made to the EICS decision-making model is shown on Fig. 2:

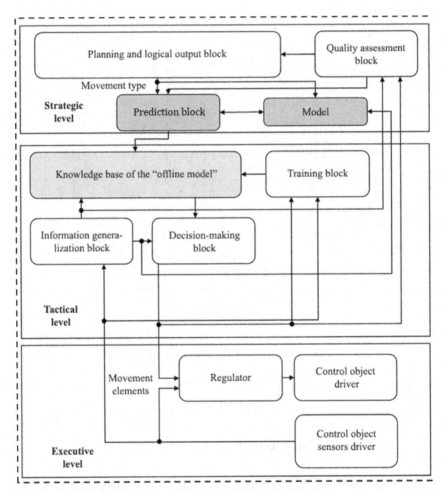

**Fig. 2.** Concept of the improved EICS model

The authors propose to make changes to the strategic level, by adding blocks "Prediction Block" and "Model" as well as making changes to the tactical level, by replacing the "Knowledge base" block with the "Knowledge base of the 'offline model'" block, in addition, the connectivity of individual blocks changes somewhat.

Due to the widespread use of simulation tools that allow us to significantly speed up the development process by conducting virtual simulations of physical objects in an easily modifiable environment, we assumed that the initial set of rules of movement and behaviour of the robot will be formed by the results of the simulation and will become the basis of the knowledge base "Knowledge base of the "offline model"". These sets of rules should serve as a starting point for the robot's continued operation and improvement. We should note that the rules laid down in this knowledge base are not unchangeable: the control system can modify these rules in the process of functioning of the robot. Also, this base is assigned an "emergency function" because if it is necessary to "reset" the information received during the operation of the base or if there are failures, the robot will be able to use the initial version of the knowledge base.

In addition, the model built in the simulation environment is ported to the robot's control system and becomes part of the strategic level (the "Model" unit) with which the "Prediction block" is in constant interaction. The essence of this interaction is as follows: "Prediction block" and "Planning and logical output block" receive information on the current state of the robot and, based on the data obtained, before performing the next action of the robot, perform a simulation "on the fly" using the model of the robot built into the control system. In this way, the robot control system, based on the information received from the sensors, will build a virtual environment that is the prototype of the real one, and place the robot model there to find a suboptimal solution for performing the robot's movement. We should note that as a result of modelling, many solutions to a specific problem can be obtained, however, the search for all possible solutions may take too much time, therefore, the search for a suboptimal solution takes place over a limited quant of time, after which the most optimal solution is chosen from a set of obtained ones. Next, the virtual environment is rebuilt according to the new information from sensors and the simulation is repeated again while the robot performs the motion command obtained at the previous step. Thus occurs predictive simulation of the movement of the robot for a variety of potential variants of control influences using the model obtained during the development phase, after which further physical execution is governed by the variant of the control influence, for which the predictive model achieves the best quality of movement (in a specified sense, for example, the length of the trajectory, the amount of energy it takes to carry out the movement, the safety of the route, etc.).

We should note that any control system cannot exist separately from the object of control. The combination of software and algorithmic solutions, which are control systems, with hardware solutions, which are control objects, allows to achieve a significant improvement in the technical and performance characteristics of robotic equipment.

The creation of fundamentally new design solutions allows to expand the ability of robotic platforms to overcome the irregularities and heterogeneity of the terrain, which must be traversed to achieve the operation goal. Recently, walking robots became widely used for solving tasks of this kind.

Interest in walking robots is due to a number of advantages of this type of design over others:

- The ability to overcome obstacles with their height up to the robot's limb attachment level, while, as the leg does not have to come into contact with the surface, some obstacles can be stepped over;
- The ability to move in any direction from the spot;
- The ability to reduce hull vibration when driving over very rough terrain.

Unlike wheeled robots, walking robots have greater manoeuvrability and better capabilities for different kinds of adaptive algorithms.

The search for new design solutions always goes hand in hand with the search for solutions in the field of control systems. The combination of advanced algorithms and design solutions is an important component of the successful and fail-safe functioning of the robotic platform.

We will consider a specific example of a new robotic platform developed: a symmetrical walking robotic platform (SWRP) proposed by the authors [13], on the basis of which the proposed concept of an embedded intelligent control system will be considered.

At the heart of this platform is a symmetrical three-dimensional propulsion system (STP). Every propulsion system (Fig. 3) contains a linear-sliding support (1) moving inside the guide attachment (2) and driven by a servo (3). The node (1,2,3) moves in the vertical plane are carried out by a servo (4) attached to the servo (5), which performs horizontal movements of the engine by a rigid joint (6). For the contact of the support with the surface, as well as to reduce pressure on the surface, at the ends of the support are platforms (7), attached to the support by hinges (8).

The robotic platform, equipped with such propulsors, has a set of basic configurations illustrated at Fig. 4 (the basic operation mode, the mode of overcoming high obstacles without changing the trajectory, the mode of increasing stability, including on the slopes, the mode of movement on the surface with irregularities, the mode of movement on surfaces with longitudinal recesses without changing the trajectory, the mode of movement on the narrow sections of the track). These structural configurations can be used both individually and in combination (for example, movement on a narrow surface with lateral unevenness is facilitated by a combination of configurations No. 6 and No. 4).

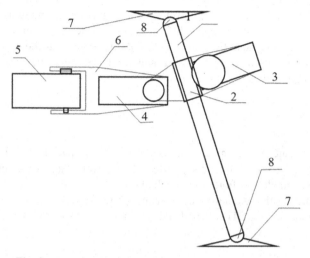

**Fig. 3.** A symmetrical three-dimensional propulsion system

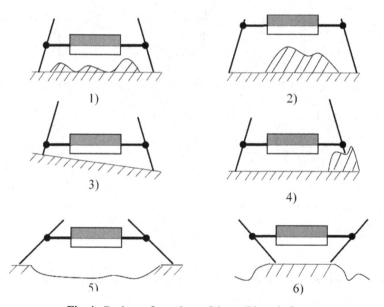

**Fig. 4.** Basic configurations of the walking platform

Due to the high pace of development of robotic systems, as well as the purpose of reducing the time and resources for development, virtual modelling tools have become widely used. Applying modelling at different stages of development can significantly reduce both the development time and the probability of error, as the developer has the opportunity to work out many different variants for designing the device even before its physical incarnation appears.

## 4 Results

The V-Rep package has been selected as the modelling package, as it has a set of advantages: it has a broad range of supported programming languages, supports a wide range of physical modelling libraries (so-called "engines"), and allows to facilitate client-server interaction with the control device with its built-in tools.

A virtual model of the STP was realized using the V-Rep simulation package (Fig. 5).

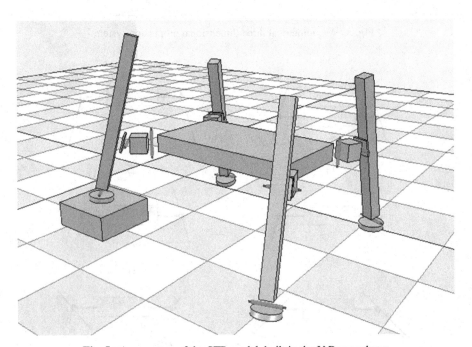

**Fig. 5.** Appearance of the STP model, built in the V-Rep package.

Then the basic configurations of the robotic platform were modelled (Fig. 6):

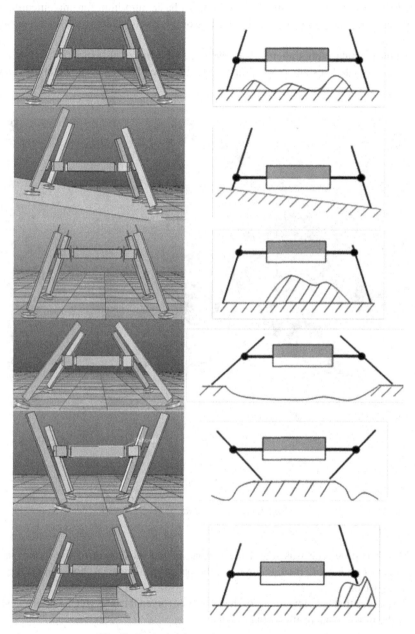

**Fig. 6.** Basic configurations of the STP model

After the implementation of the model of the robotic platform in the simulation environment, the execution of movements by the executive level of the robotic platform is modeled. The purpose of this stage is to find a basic algorithm for implementing the gait of a robotic platform. A real biological prototype was taken as a basis - a monitor lizard.

For the physical embodiment of the developed STP, first of all, a model was developed using an automated design system. Further, a prototype of the STP, shown in Fig. 7, was built.

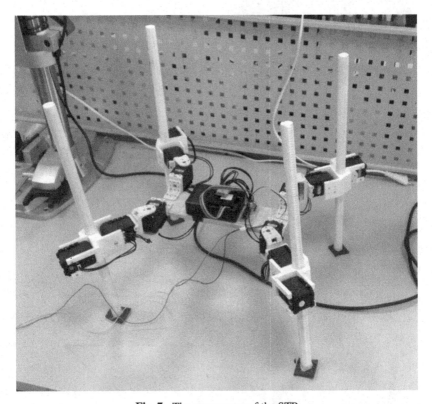

**Fig. 7.** The appearance of the STP

The STP prototype was tested by comparing the movement of the model in the virtual environment and the movement of the prototype. Figure 8 shows an example of this comparison.

We can see that the movements of the prototype (given the imperfection of the design) repeat the movements of the virtual model.

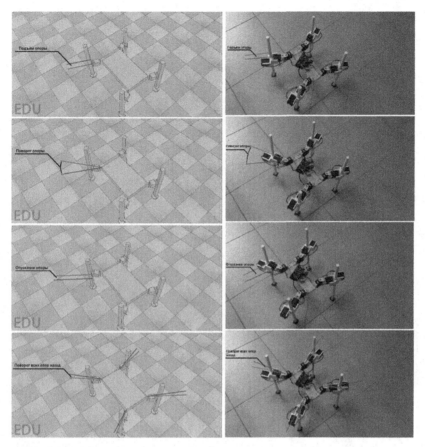

**Fig. 8.** Comparison of the movement of the virtual model and the real prototype

To control the robot model, as well as a real prototype, a control software package was developed based on the Robot Operating System. The structure of the developed system is shown in Fig. 9.

The proposed control system is universal and can be widely used in robotics.

In particular, a robot equipped with a similar control system can perform a virtual simulation of its movement in the terrain in advance, saving and analyzing certain metrics of its movement and choosing the most optimal path according to a certain criterion.

At the same time, it is possible to work in several modes at once: with or without a complete map of the area.

In the absence of a terrain map, it is possible to simulate shorter time intervals, solving the problem of local path planning. If you have a complete and detailed map of the area, you can plot a global route in advance, as mentioned above.

At the same time, it is worth noting an important detail: all these capabilities are performed autonomously based on the control system of the robot itself.

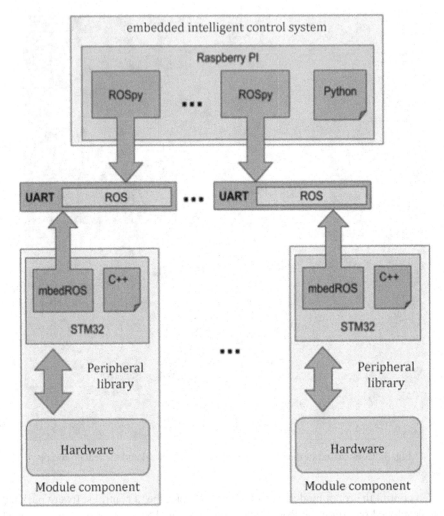

**Fig. 9.** The scheme of interaction of the components of the control system

Unlike traditional robots such as the BigDog, Cheetah, LS3, Spot, which use a biological approach (based on motion kinematics similar to that of biological species), the design of the STP allows it to move both directly and upside down, while the mentioned robots lose their functionality partially or completely when flipped. In addition, the structural features of the STP allow to implement movement on narrow and wide tracks.

## 5   Conclusion

The combination of proposed solutions can reduce labor intensity and improve the quality of the control systems being developed. In addition, the use of presented solutions will increase the potential of adaptive control algorithms, shorten the development time of robotic complexes, as well as significantly improve the ability of robotic complexes to overcome irregularities and heterogeneity of the terrain.

The proposed EICS model can find its application in a wide range of modern robotic complexes, as the approaches formulated within the proposed concept are universal and do not depend on the type (wheeled, walking, floating, etc.) or on the approach to building a robotic complex.

The proposed design of the STP differs in a number of features (the ability to function in an overturned state, the ability to move on narrow or wide tracks, the ability to dynamically control its clearance, etc.), confirmed by the patent for the invention, can significantly improve the ability to overcome irregularities and heterogeneity of the terrain, as well as improve the fault tolerance of the robotic complex during search and rescue operations.

In the future, it is planned to implement the tools of transferring the virtual model to the EICS, carry out the transfer of the virtual model to the EICS, conduct tests using all sorts of robotic complexes and log their functional characteristics.

## References

1. Denker, A., İşeri, M.C.: Design and implementation of a semi-autonomous mobile search and rescue robot: SALVOR. In: 2017 International Artificial Intelligence and Data Processing Symposium (IDAP), pp. 1–6 (2017). https://doi.org/10.1109/IDAP.2017.8090184
2. Vassiliev, A., et al.: Designing the built-in microcontroller control systems of executive robotic devices using the digital twins technology. In: Proceedings of 2019 International Conference on Information Management and Technology (ICIMTech-2019). INSPEC Accession Number: 18995221. https://doi.org/10.1109/ICIMTech.2019.8843814
3. Aoki, T., Asami, K., Ito, S., Waki, S.: Development of quadruped walking robot with spherical shell: improvement of climbing over a step. ROBOMECH J. **7**(1), 1–12 (2020). https://doi.org/10.1186/s40648-020-00170-5
4. Boston Dynamics Shows Latest Advancements in Robotics at CEBIT 2018. - YouTube. https://www.youtube.com/watch?v=GZgNPmeaTjo. Accessed 18 Nov 2018
5. Dissanayake, M., Sattar, T.P., Howlader, O., Pinson, I., Gan, T.H.: Tracked-wheel crawler robotfor vertically aligned mooring chain climbing design, simulation and validation of a climbing robot for mooring chains. In: 2017 IEEE Inter-national Conference on Industrial and InformationSystems (ICIIS), pp. 1–6 (2017)
6. Hu, N., Li, S., Gao, F.: Multi-objective hierarchical optimal control for quadruped rescue robot. Int. J. Control Autom. Syst. **16**(4), 1866–1877 (2018). https://doi.org/10.1007/s12555-016-0798-8
7. Moskvin, I., Lavrenov, R., Magid, E., Svinin, M.: Modelling a crawler robot using wheels as pseudo-tracks: model complexity vs performance. In: 2020 IEEE 7th International Conference on Industrial Engineering and Applications (ICIEA), pp. 1–5 (2020). https://doi.org/10.1109/ICIEA49774.2020.9102110
8. Jianye, N., et al.: Study on structural modeling and kinematics analysis of a novel wheel-legged rescue robot. Int. J. Adv. Rob. Syst. **15**, 172988141775275 (2018). https://doi.org/10.1177/1729881417752758
9. Pecka, M., Zimmermann, K., Svoboda, T.: Fast simulation of vehicles with non-deformable tracks. In: 2017 IEEE/RSJ International Conference on Intelligent Robots and Systems (IROS), pp. 6414–6419 (2017). https://doi.org/10.1109/IROS.2017.8206546

10. Micire, M.J.: Evolution and field performance of a rescue robot. J. Field Robot. **25**(1–2), 17–30 (2008)
11. Rouzbeh, B., Bone, G.M.: Optimal force allocation and position control of hybrid pneumatic-electric linear actuators. Actuators **9**(3) (2020). https://doi.org/10.3390/act9030086
12. This robotics startup wants to be the boston dynamics of China. - IEEE Spectrum. https://spectrum.ieee.org/automaton/robotics/robotics-hardware/this-robotics-startup-wants-to-be-the-boston-dynamics-of-china. Accessed 1 Nov 2020
13. Vasilev, A.E., Vasilianov, G.S.: Symmetrical three-dimensional propulsion system with a linear-sliding support and embedded control device, and a symmetrical walking platform based on it. Russian Patent for Invention No. 2643613, published 10.03.2016 bulletin 2016108738
14. Wang, Y., Tian, P., Zhou, Y., Chen, Q.: The encountered problems and solutions in the development of coal mine rescue robot. J. Robot. **2018**, 8471503, 11p (2018) https://doi.org/10.1155/2018/8471503
15. Yoshida, T., Nagatani, K., Tadokoro, S., Nishimura, T., Koyanagi, E.: Improvements to the rescue robot quince toward future indoor surveillance missions in the Fukushima Daiichi nuclear power plant. In: Yoshida, K., Tadokoro, S. (eds.) Field and Service Robotics. Springer Tracts in Advanced Robotics, vol. 92. Springer, Berlin (2014). https://doi.org/10.1007/978-3-642-40686-7_2
16. Zhang, K., Gao, R., Zhang, J.: Research on trajectory tracking and obstacle avoidance of nonholonomic mobile robots in a dynamic environment. Robotics **9**(3) (2020). https://doi.org/10.3390/ROBOTICS9030074

# Author Index

Printed in the United States
by Baker & Taylor Publisher Services